MW01639860

Guide to Fragrance Ingredients

The H&R Book

Guide to Fragrance Ingredients

Johnson · London

Authors
Rüdiger Hall, Dieter Klemme, Dr. Jürgen Nienhaus, Holzminden,
West Germany

With Cooperation of
Dr. Ulrich Harder, Dr. Rudolf Hopp, Klaus Peters,
Werner Reifenstein, Bernfried Warnecke, Holzminden, West Germany

Concept
Lothar F. Kümper, Karl-Heinz Schmid, Holzminden, West Germany

Design/Graphics
Fred Niemüller, Ronnenberg/Benthe, West Germany

Coordination, Production
Bernhard Kott, Siegfried Rogas, Holzminden, West Germany

Lithography
Hilpert, Essen, West Germany
Jahn Repro, Bielefeld, West Germany
Lithografie Service, Garching, West Germany

Printing
Frohne Druck, Bad Salzuflen, West Germany

English edition published by
Johnson Publications Limited
130 Wigmore Street, London WIH OAT

Printed in Germany
ISBN 3-87261-053-8

Contents

Volume 1

The H&R Book of Perfume

How is a perfume produced? What is a perfume made of? Where do the raw materials come from and how are they extracted? What are the psychological aspects of fragrance? How is perfume effectively stored? What are the most effective ways to use perfume? The answers to these and many other questions are to be found in "The H&R Book of Perfume".

160 pages with 172 color photographs and 35 color illustrations.
Dimensions 9 ½ x 12 inches, Hardcover

Volume 2

Fragrance Guide Feminine Notes

Introduction to the classification of fragrances according to families. Color illustrations of 421 original perfume bottles presently on the international market, classified according to fragrance families. Each product with fragrance description according to top, middle and base notes. Information on the type-determining ingredients of each perfume, as well as the producer of the perfume.

144 pages with 435 color photographs and 1 multicolored illustration.
Dimensions 9 ½ x 12 inches, Hardcover

Volume 3

Fragrance Guide Masculine Notes

Presentation of 361 products on the international market, including corresponding descriptions of the perfumes and classification following the same pattern set by the "Fragrance Guide Feminine Notes".

144 pages with 381 color photographs and 1 multicolored illustration.
Dimensions 9 ½ x 12 inches, Hardcover

Volume 4

Guide to Fragrance Ingredients

A volume which defines 341 natural and synthetic fragrance ingredients, according to the place of origin, production and fragrance in an easy-to-understand and factual manner. The processes of production of aroma chemicals are described in detail and graphically reproduced.

144 pages with 124 color illustrations and 17 multicolored graphics.
Dimensions 9 ½ x 12 inches, Hardcover

Foreword

With the rich abundance of plants she offers us, Nature serves as a huge reservoir of odorants. And, in addition to Nature's flora, her fauna, too, is of invaluable significance to perfumery. The same also holds true with respect to the wealth of synthetic aroma chemicals without which the fragrance organ played by perfumers today would be inconceivable.

With the H&R Guide to Fragrance Ingredients, a comprehensive reference work is now available, in which 341 fragrance elements of natural and synthetic origin are clearly and illustratively categorized. In additon to 141 full-color illustrations, it depicts the botanical origins or chemical formulae of these fragrance elements, as well as how they are obtained, their fragrances and their applications. Information that is of interest not only to those in the trade, but also to anyone who uses and enjoys perfumes and fragrances.

The methods employed in obtaining natural fragrance elements and producing synthetic aroma chemicals provide a fascinating insight into a branch of the chemicals industry that makes our daily lives a bit more "fragrant".

This book serves as a meaningful complement to Volume I, the H&R Book of Perfume, as well as Volumes II, H&R Fragrance Guide Feminine Notes, and III, H&R Fragrance Guide Masculine Notes. The complete edition treats all questions pertaining to the broad spectrum that makes up the world of perfume: 800 of the most prominent brand-name perfumes are classified, and a perfumistic description is provided for each.

Without the support and counsel of the experts from the house of Haarmann & Reimer in Holzminden, Germany, publication of this comprehensive, definitive work would not have been possible. To all of the firm's employees who so willingly gave of their time and know-how go my sincere thanks.

Wolfgang Glöss,
Publisher

Nature – Supplier of Fragrant Perfume Components

For thousands of years man has been utilising the riches of nature in order to make his life and environment more fragrant. Amongst the flora of this earth there is a multiplicity of plants that have the capacity of retaining and storing aromatic substances generated by their metabolism. The essential oils, indispensable in perfumery, are a good example. And man has learned to augment the natural supply by cultivating the aromatic plants on an ever-increasing scale.

Where and How does a Plant store its Essential Oil?

Aromatic substances are formed and stored in certain organs of a plant as by-products or indeed the end result of its metabolism. Amongst the various parts of a plant that can contain aromatic materials there are the leaves, flowers or fruits; the wood of limbs or tree-trunk, and the roots.

Here are some examples:
Glandular Cells, Glandular Hairs and Glandular Scales. These are single or multi-cell protuberances, or "pockets," on the surface of the plant's epidermis. Thyme, marjoram, bean-plant, rosemary and sage, all of the family Lamiaceae (Labiates) are representatives of the group that stores essential oils in this manner.

Oil Cells and Resin Cells. These are cells, still living in some cases, which are filled with oil or resin in plants of the laurel family (Lauraceae), laurel leaves, cinnamon and cassia being well-known examples.

Oil or Resin Canals. Inter-cellular spaces in plant tissue store essential oils and resins. When adjoining glandular cells move apart from each other, the spaces expand into tubular ducts. Canals or ducts formed in this way are found in the schizocarp fruits of the Apiaceae (Umbellifers), e.g. caraway, aniseed, fennel, coriander and celery. The conifers, too, have their resin canals. If the tree is damaged, large quantities of resin can be exuded. Some resins are, in fact, systematically gathered by this method of "tapping."

Oil Reservoirs. Lysigenous secretory reservoirs are formed inside a plant as the walls of secretory cells gradually disintegrate. This is called secondary cavity formation. The rue family (Rutaceae), in particular, is noted for such lysigenous cavities, or oil reservoirs; citrus varieties such as lemon, orange and bergamot are outstanding examples.

Why are Many Different Processes Needed to Obtain Aromatic Substances?

The choice of process depends on the nature of the material to be worked and/or the characteristics one seeks in the end-product. The basic methods are pressing (expression), extraction and distillation, each with many variations and refinements.

A very mild and gentle treatment, used exclusively for obtaining citrus oils, is expression (at ambient temperature). Citrus oils are easily damaged if subjected to heat which causes changes in their chemical structure and ipso facto to the odor-value–a telling reason for using the "cold-press" method.

The one exception among the citrus family is lime oil. In this case the change induced by the heat of distillation is actually desirable as it gives the odor a different character.

Extraction is the most suitable process, as regards economy and quality of end-product, for materials with a very small content of volatile aromatic substances–jasmin blossoms, for example. It is also very suitable for raw materials, such as resins which contain "heavy" aromatic matter of low volatility. In the case of jasmin, extraction yields first the so-called concrète which is then processed further (second extraction) to the "absolute." The yield from resins is called a resinoid.

Like expression, the extraction process is mild and gentle, as witness the fact that essential oils obtained by this means generally have finer, better odors than their counterparts produced by steam distillation. One drawback of the extraction method is that it brings out, in addition to the volatile odorous substances, firstly components of the plant that are practically non-volatile and therefore have fixative powers whilst being virtually odor and colorless, and secondly vegetable coloring matter such as chlorophyll. For this reason, concrètes, absolutes and resinoids are all colored, to a greater or lesser degree; also, unlike the essential oils, they do not evaporate completely without leaving a residue.

The main advantage of distillation is that it can be carried out easily, economically and, in some cases, with very simple apparatus at the locations where the oil-bearing plants are harvested or collected.

Compared with extraction methods, distillation is less labor-intensive. Large quantities of material can be processed in relatively short periods of time without incurring the hazards of handling big volumes of highly inflammable solvents. A disadvantage of distillation has already been mentioned: odors of essential oils thus produced are liable to modification by the effects of heat and water.

Methods of Producing Natural Fragrance Materials.

Expression

This process is used only for obtaining oils from the citrus fruits.

The outer layer of the peel is ruptured by mechanical means and the oil is then pressed out. In this manner, the so-called "cold-pressed" grades of orange, lemon, bergamot, mandarin, grapefruit and lime oils are obtained. Lime oil can also be distilled, thereby acquiring a distinctively different aroma.

Extraction

is a procedure by which aromatic substances are drawn or leached out of the raw material by means of solvents. These latter are selected individually with regard to the character and composition of each raw material. There are many liquid solvents used by industry, especially petroleum ether, hexane, toluene, methanol and ethanol. There are also gases such as butane and carbon dioxide which become liquid under pressure.

Animal fats are used as "solvents" in the old *Enfleurage* process, rarely used today, and only for extraction of flower oils. *Enfleurage à froid* means placing freshly picked blossoms on plates of glass that have been coated on both sides with animal fat–which absorbs the fragrance. When the blossoms are exhausted and fade, they are replaced by fresh flowers and these renewals continue until saturation point is reached. Fat thus saturated with flower scent is called *Pomade.*

From this pomade, the fragrant substances are "washed out," i.e. liberated with alcohol. The alcohol is then evaporated, leaving as a residue the end-product: *Essence absolue de Pommade,* or absolute ex pomade.

If the fat is heated and melted before being used as a "solvent," the process is called *Enfleurage à Chaud.*

As already mentioned, both processes are hardly used these days as they are inordinately wasteful and altogether too expensive.

The extraction method in general use today requires non-polar solvents such as petroleum ether, hexane or toluene. Fig. 1 shows diagrammatically a solvent extraction process for producing *Concrètes.*

The material to be processed is placed in an extraction vessel, there immersed in a non-polar solvent and heated gently. After a pre-determined interval, the solvent which has absorbed aromatic substances is replaced by a fresh batch. The choice of solvent, the frequency of solvent renewal, the working temperature and the duration of the whole process all vary from product to product.

Separation of the end-product from the solvent is a two-stage affair. Phase I takes place in a large still at normal pressure or low vacuum, and phase II in a smaller still under high vacuum. The recovered solvent is used again for further extractions; the solvent-free residue of aromatic substances is called *Concrète.*

Conversion of Concrète to Absolute

Fig. 2 shows, again as a diagram, the preparation of an *absolute* from the concrète. The aim is to remove those parts of the concrète, meaning the waxes, which are not soluble in alcohol; in other words, to convert the concrète (itself only partially soluble in alcohol) into a fully soluble material which is easier to use as a perfume ingredient.

The first step is to mix the concrète with alcohol, heating very carefully to approx. 50 °C. Even at this temperature part of the wax may not dissolve. The mixture is allowed to cool, slowly, to below 5 °C., and is then filtered. Removal of the alcohol from the resulting clear, wax-free solution is similar to the treatment of concrètes. i.e. first at normal pressure or weak vacuum, and secondly under high vacuum.

Recovered alcohol can be used again, either for the next batch or for a second treatment of waxes that have been filtered out. There remains the end-product: the alcohol-free but clearly soluble absolute.

Manufacture of Resinoids
Fig. 3 illustrates the production of *resinoids.* The raw materials are usually solids or pastes–resins, in fact. The aim is to extract from the resin not only the odorous material but also the virtually odorless substances which have fixative properties because they are practically non-volatile. Extraction of resins is carried out in vessels fitted with stirring mechanism (agitators) using solvents such as toluene, chlorhydrocarbons or alcohol at varying temperatures. The resulting solution is then filtered. Afterwards, the solvent is driven off in a still, first at normal pressure or under low vacuum, and finally–to remove the last vestiges of solvent–under higher vacuum appropriate to each individual product. Recovered solvent is fed back into the next extraction.

The solvent-free end-product–the resinoid–is sometimes thinned with virtually odorless, low-volatile diluents such as diethyl phthalate or benzyl benzoate merely to achieve a more liquid consistency which will facilitate handling.

Extraction with Carbon Dioxide (CO_2)
Figs. 4 and 5, showing the physical states and the procedures in this modern extraction technique are self-explanatory. The main advantage of this process is its extreme gentleness which is of great value in the production of delicate natural fragrance materials that are sensitive to heat. CO_2 can be used either in liquid form at temperatures below 31 °C, and at appropriate pressure, or at temperatures above the critical 31 °C, at pressures of more than 75 bar. The extractive powers of carbon dioxide, and consequently the yield and the quality of the product, vary with the parameters.

Distillation
The distillation method, in other words evaporation and subsequent condensation of liquids, is applied–with numerous variations of detail–to the production, rectification and concentration of natural aromatic materials.

Examples are given hereunder:

Steam Distillation
As Fig. 6 shows, both "direct" and "indirect" steam distillation can be used.
In the case of the *direct method,* most commonly employed for producing essential oils, the still is charged with the material to be processed. Steam is then introduced at the base of the still over a distributor plate. The volatile elements of the material mix with the steam, and this vapor-mix separates, after condensation to liquid form, into water and essential oil. Continuous separation of oil from water is effected in a "Florentine Flask" from which the water–greatly superior in volume to the oil–is run off through a so-called swan-neck (siphon).

In the case of *indirect* steam distillation–*water distillation* or boiling, to be precise–the still is filled with both water and distillation material and brought to the boil. As in the case of the direct method, the steam combines into a vapor mixture with the volatile components of the aromatic substance. Water from the Florentine Flask is fed back to the still (cohobation).

Indirect steam distillation is seldom used these days as it requires more energy for a longer processing time and gives a lower yield of oil into the bargain. Fig. 7 shows an example of indirect steam distillation producing orris root oil.

Hydro-Diffusion (see Fig. 8)
could be described as a variation on direct steam distillation. The main difference is that the steam is introduced not at the base of the still but at the top, i.e. lid-level. Condensation of the oil/steam mixture occurs inside the still, directly below the grill or perforated tray supporting the material being processed. Lower steam consumption, shorter processing time and increased oil yields are said to be the advantages of this process over the conventional direct steam distillation method.

Vacuum Distillation
This process is used to modify essential oils

already produced by steam distillation. The aim could be to improve both concentration and solubility of the oil by total or partial elimination of terpene and sequi-terpene carbohydrates. Also, vacuum distillation can increase the volume of principal components in the oil, or even isolate such constituents. At the same time, one finds that certain oils which are dark in color (owing to their nature and the method of manufacture, or because of prolonged storage) become paler under vacuum distillation.

In contrast to steam distillation the vacuum process is a *dry distillation,* meaning that the steam does not enter the still itself but only a surrounding heating jacket. Mounted on the still is a column (height determined by nature of the work in hand) filled with suitable "packing." The components of the oil being distilled will evaporate at different times, each at its own boiling point. In the column, which serves to separate accurately the various components (or groups of components), there is an exchange of rising product-vapor with the condensate running back to the still.

Vacuum causes the whole oil to boil at lower temperature. The higher the vacuum, the lower the boiling point and, ipso facto, the risk of thermal degradation in the oil. The vaporised product is liquefied in condensers, and cooled, and the resulting product collected in distillation flasks. Alternatively, one may decide to catch, according to a pre-determined plan, the various portions = *fractions* of the distillate as they emerge from the condenser. Differing from one another, as they will, in odorvalue, boiling point, etc., each of these fractions can be regarded as a perfume component in its own right.

Fig. 9 shows a conventional vacuum distillation arrangement.

Molecular Distillation
is used to cope with particularly difficult distillation problems, i.e. materials which have very high boiling points and either cannot be processed by normal methods or would suffer damage by heat. Molecular distillation, illustrated by Fig. 10, is an extreme measure in every sense. The vacuum is very high (up to 7 x 10^{-5}mbar) and the distance between the (thin-layer) evaporator and the condenser extremely small. Molecular distillation is used for natural aromatic substances, for instance to remove or reduce the color of extracts which are very dark by nature.

In addition to those mentioned above, there are numerous other methods of producing natural fragrance materials. Consequently, this review does not claim to be exhaustive, nor to be accurate in every detail. It is intended solely to give an insight into the complexities of natural fragrance material production.

Solvent extraction unit

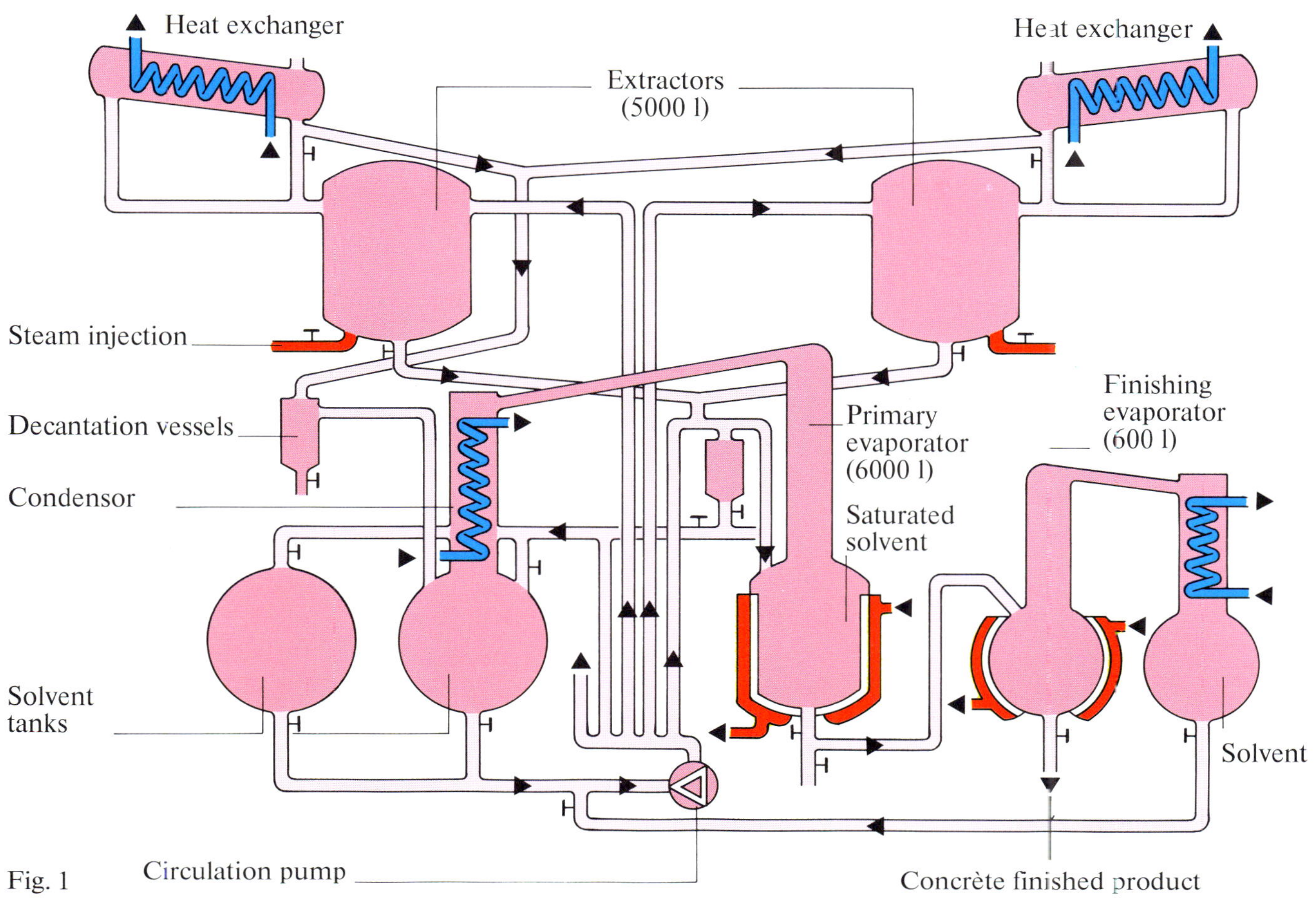

Fig. 1

Transformation of concrètes into absolutes

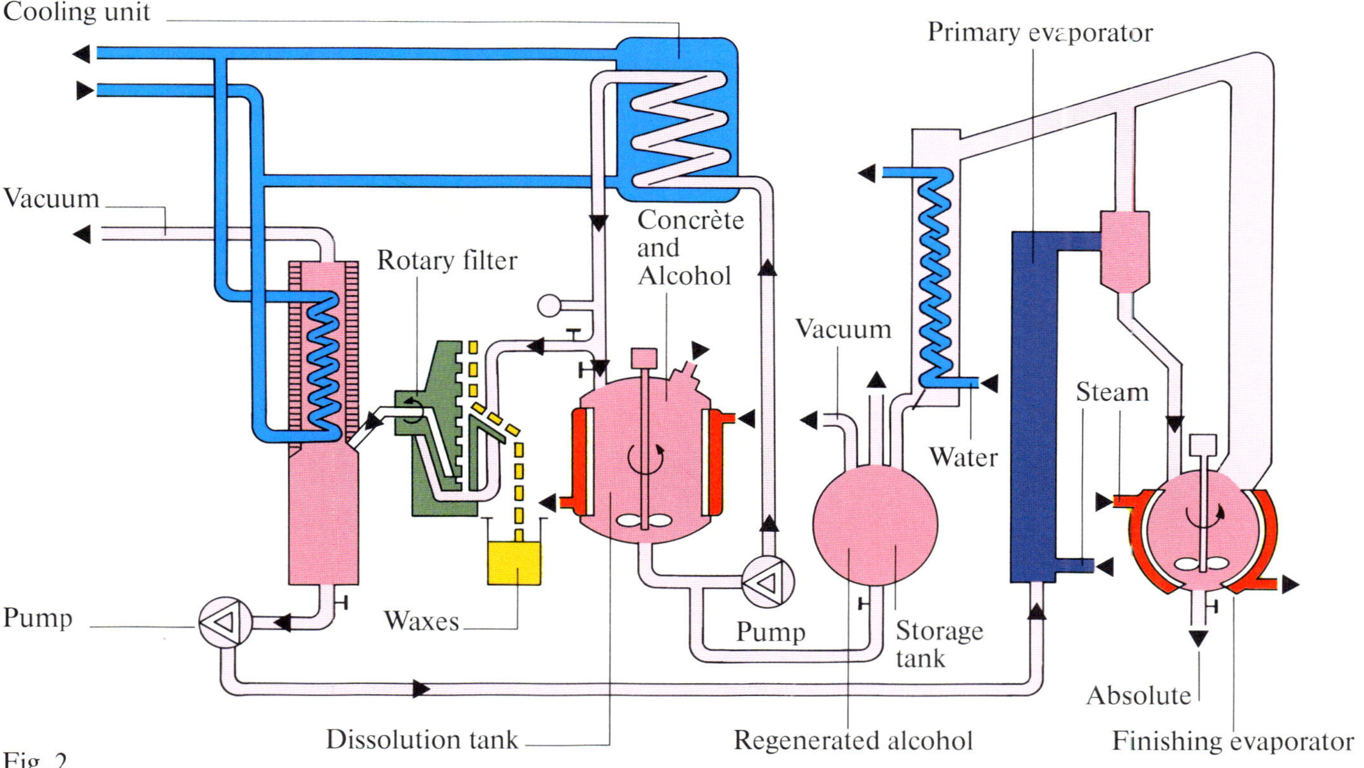

Fig. 2

Resinoid extraction unit

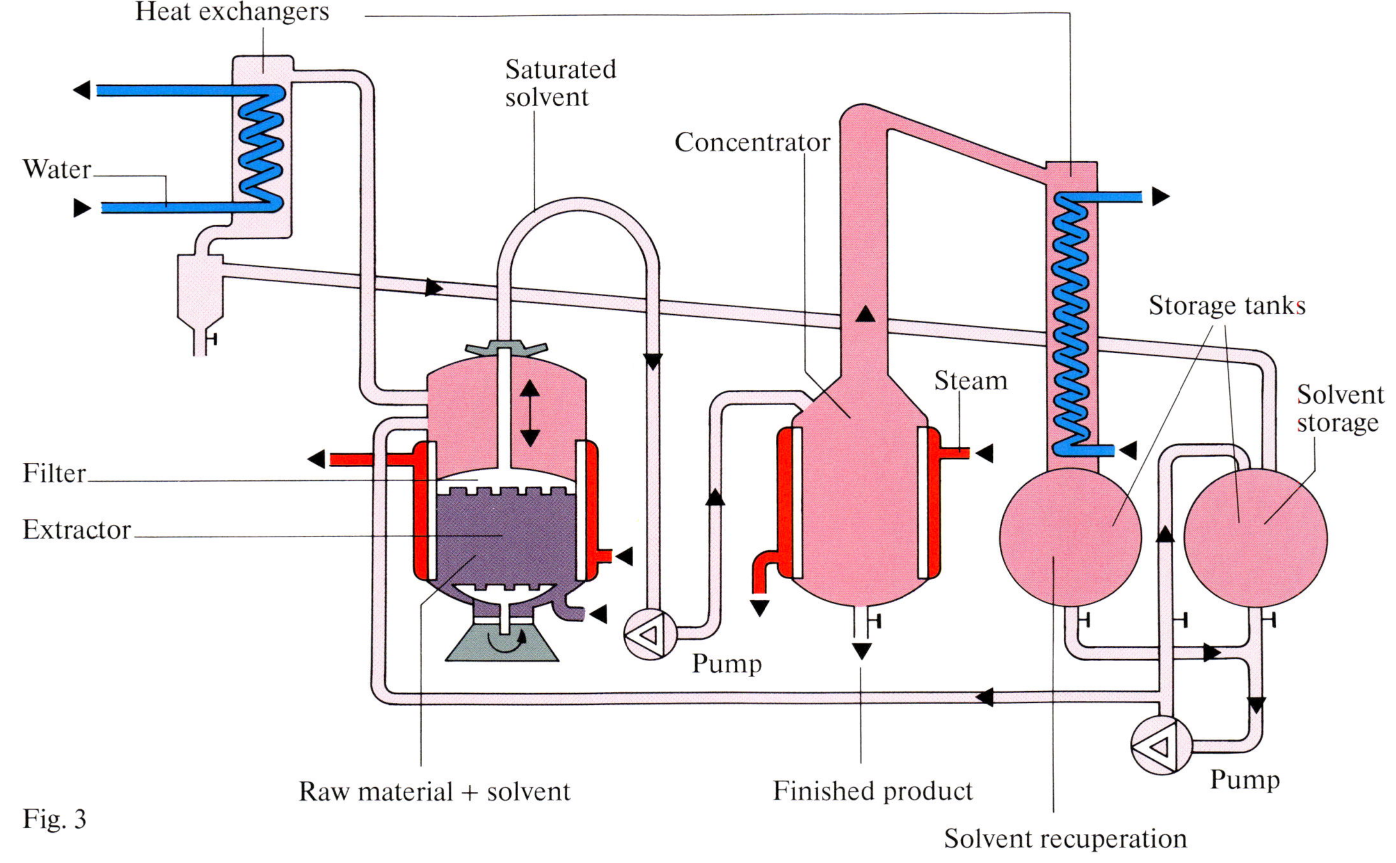

Fig. 3

CO_2 Extraction diagrams – physical states and process

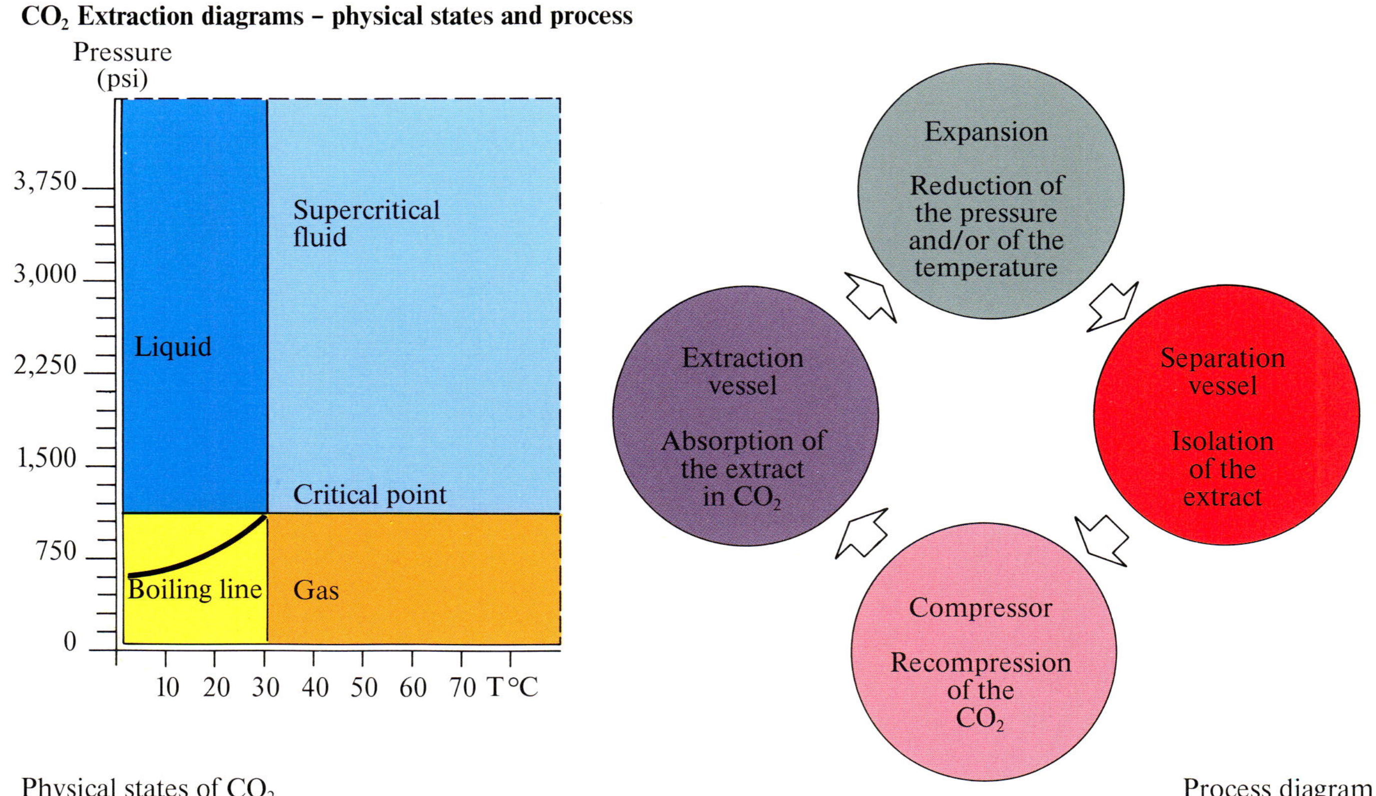

Physical states of CO_2

Process diagram

Fig. 4

CO_2 Extraction diagram

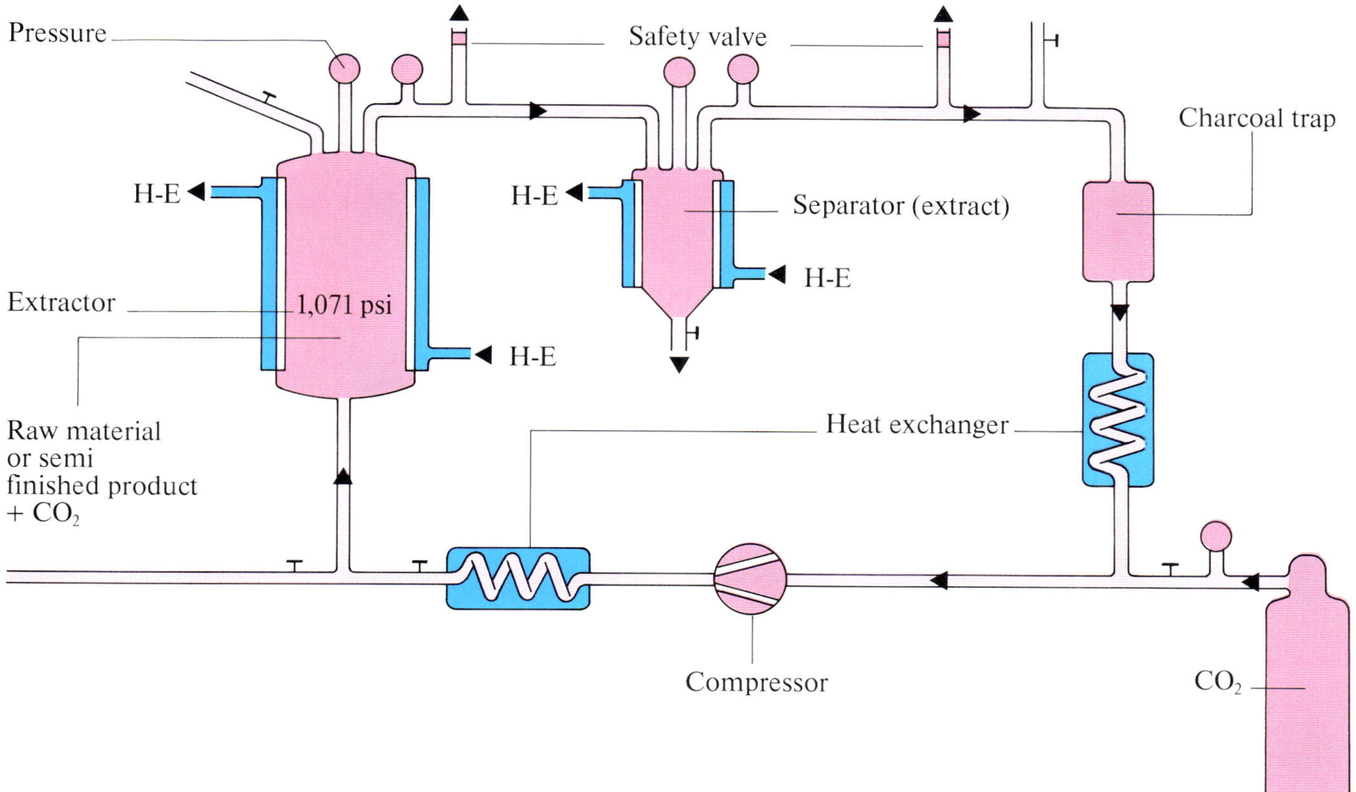

Fig. 5

Steam distillation diagram

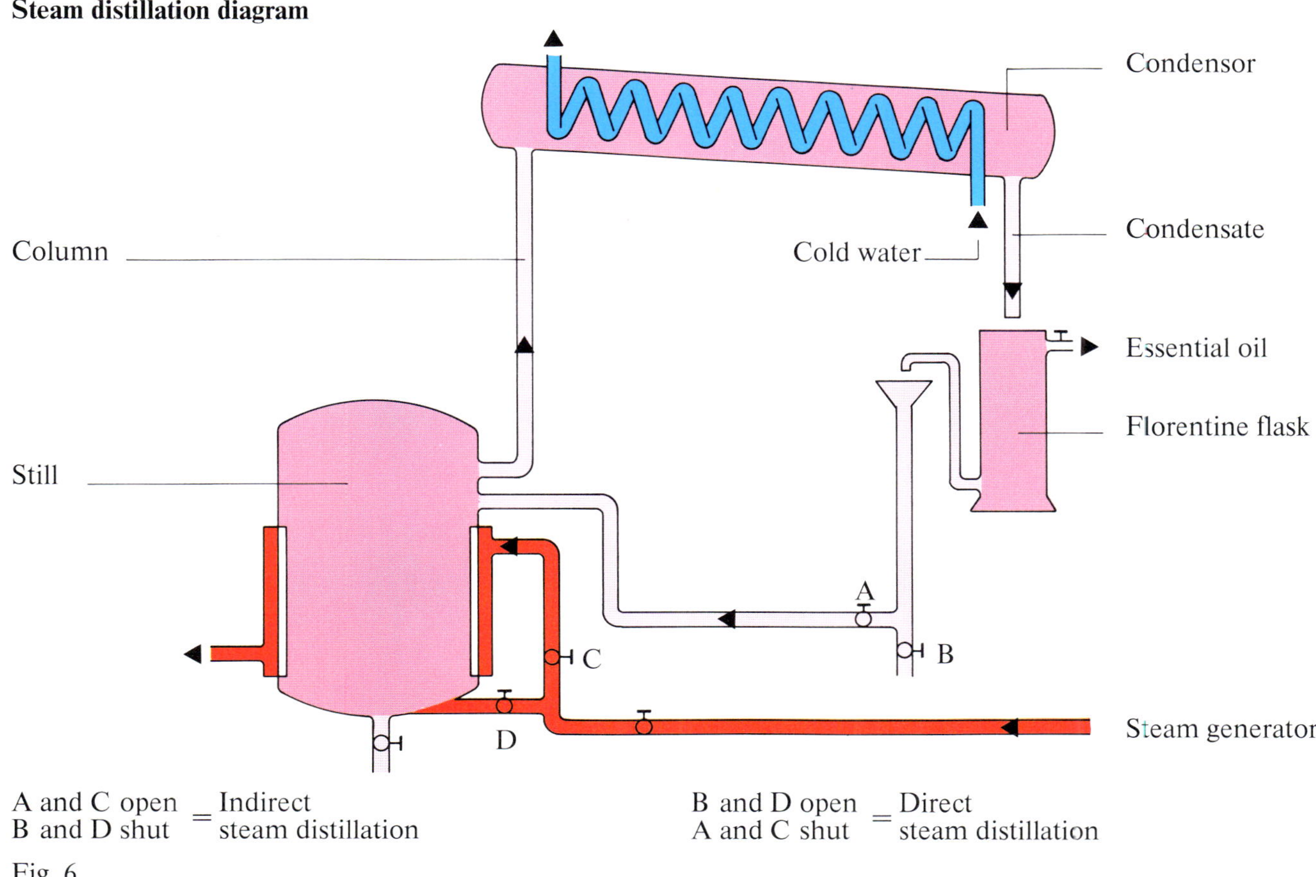

A and C open = Indirect
B and D shut steam distillation

B and D open = Direct
A and C shut steam distillation

Fig. 6

Orris rhizomes distillation

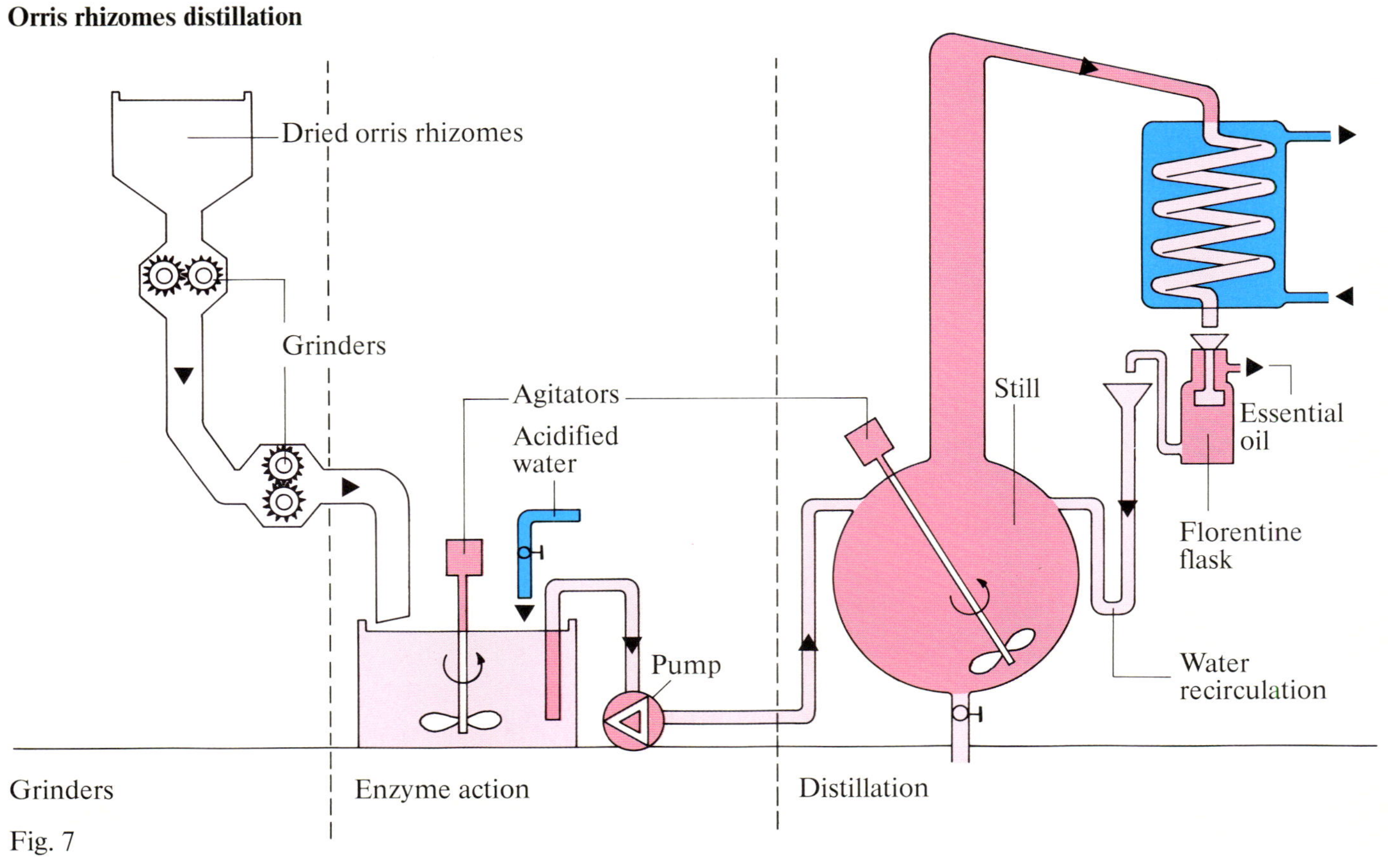

Fig. 7

Hydrodiffusion diagram

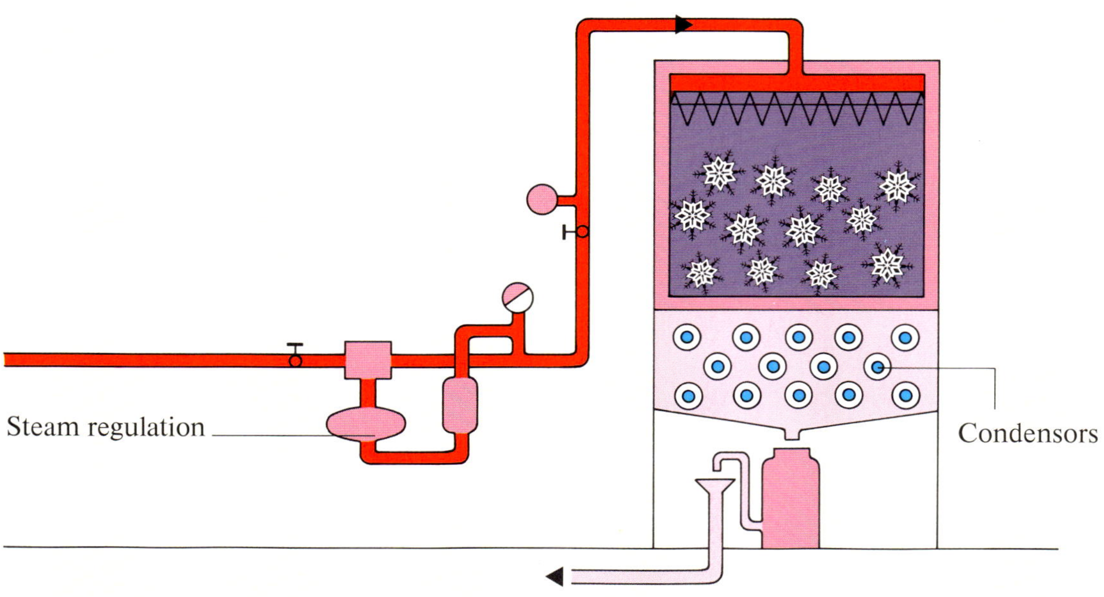

Fig. 8

Vacuum distillation diagram

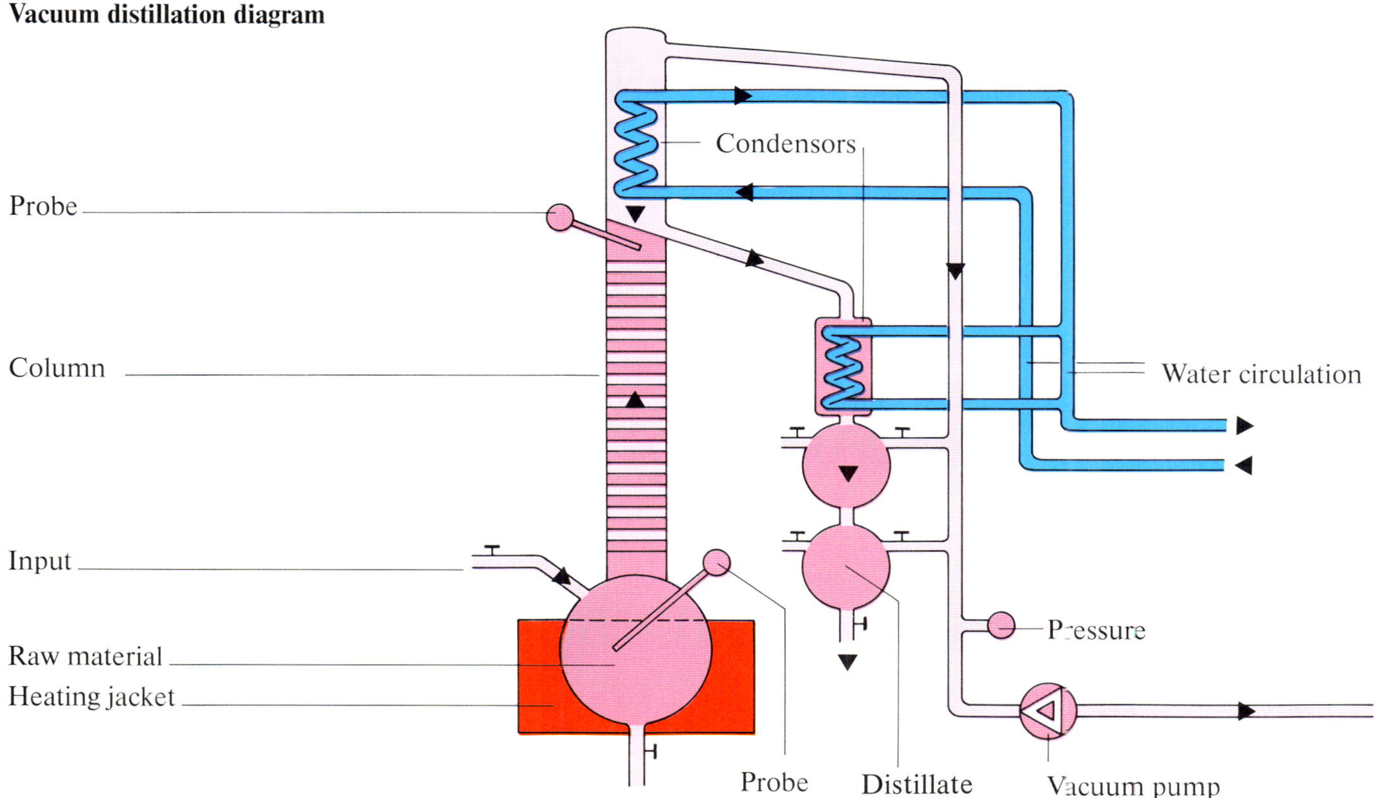

Fig. 9

Molecular distillation unit

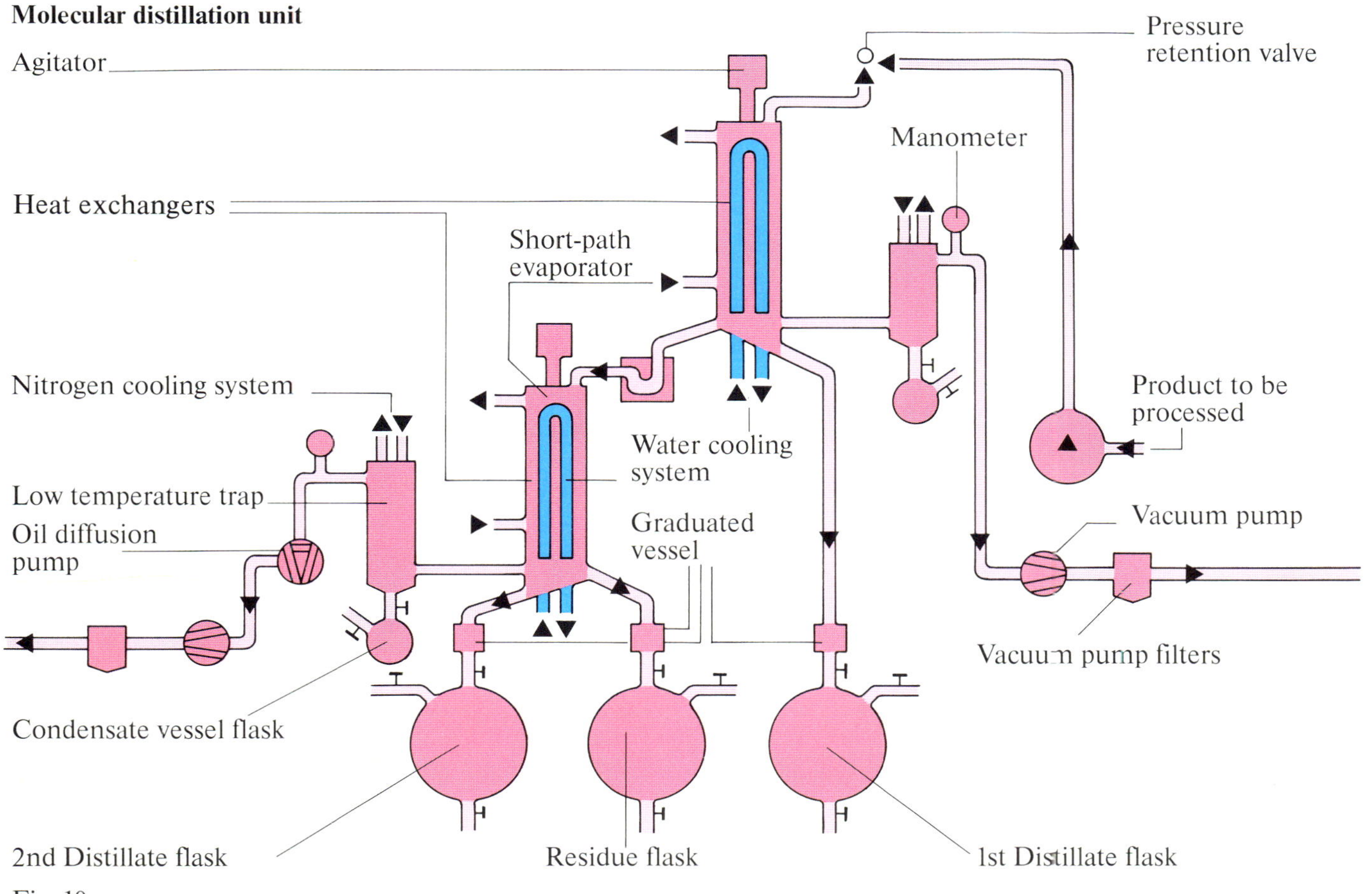

Fig. 10

Flowers from the Retort

The Production of Synthetic Fragrance Components

No naturally occuring scent goes straight into the perfume bottle. All the processes of fragrance extraction are complex. The supply of raw materials is also difficult. Land available for cultivation is becoming more and more limited as the world's population grows. The plundering of the tropical forest, the decline of natural flora through agricultural monocultures, other influences of encroaching civilization, and hunting have considerably reduced the world's supply of fragrance raw materials.

Even though the civet cat is now especially bred and fragrance plants have been cultivated for centuries, the natural raw materials available would cover only a third of the world's needs. As a result, synthetic fragrance materials have become more and more important. The synthesis of fragrance components is not restricted to the imitation of natural odors. Substances have been created in the laboratory which do not exist in nature at all. Among these are well-known fragrance components such as hydroxycitronellal, which has a beautiful, floral effect reminiscent of lily of the valley and lilac.

Furthermore, certain perfume effects, which could never have been achieved with natural fragrance materials, are possible with synthetics. These have shaped complete fragrance directions, such as the aldehyde notes (Chanel N°5) with the fatty aldehydes; or the floral notes, with benzyl salicylate (L'air du Temps). Two further examples: light, modern perfumes such as Diorella could not have been created without methyl dihydro jasmonate; and chypre perfumes such as Cabochard could not have been created without isobutyl quinoline.

The use of fragrant materials from plants and the animal kingdom, first for religious purposes, but later, increasingly for personal use, dates back to antiquity.

Compared to that, the history of synthetic fragrance substances is still very short. It began in the first half of the last century and ran parallel to the rapid development of organic chemistry. Even in those days, it was a fascinating job for the research chemists to trace the "fragrance principles" of natural perfume raw materials–in other words, to isolate them from the more or less complex natural mixtures in which they occur, and to elucidate their chemical structures.

Here are a few milestones in the development of synthetic fragrance materials:

1834 Isolation of cinnamic aldehyde from cinnamon oil (Dumas, Péligot)

1837 Isolation of benzaldehyde from bitter almond oil (Liebig, Wöhler)

1855 Synthesis of aliphatic aldehydes (Piria)

1856 Synthesis of cinnamic aldehyde (Chiozza)

1862 Synthesis of benzaldehyde (Cahours)

1874 Synthesis of vanillin from coniferin (Haarmann, Tiemann)

1875 Synthesis of coumarin (Perkin)

1876 Synthesis of vanillin from eugenol or guaiacol (Reimer, Tiemann)

1893 Synthesis of ionone (Krüger, Tiemann)

1906 Isolation of muscone from glands of the musk deer (Walbaum)

1926 Identification of the muscone structure (Ruzicka)

1934 Synthesis of muscone (Weber, Ziegler)

The systematic investigation of odor-determining main ingredients of natural fragrance materials has in many cases initiated the industrial production of important synthetic fragrance materials. In recent times, the interest of the fragrance chemist has been directed increasingly towards the fragrance-intensive by-products and trace components which–even in small concentrations–produce a characteristic effect and are therefore especially valuable to the perfumer.

Classification of Aroma Chemicals

Aroma chemicals can be produced from natural as well as synthetic raw materials, and are divided into the following groups:

Isolates. These are obtained from essential oils or other natural materials, with the help of physical or chemical separation methods, for example, eugenol from clove leaf oil, eucalyptol from eucalyptus oil, cedrol from cedarwood oil, citral from lemongrass oil, and menthol from peppermint oil. Eugenol, eucalyptol and cedrol are only available as isolates. Citral and l-menthol are also synthetically produced.

Semi-synthetic Aroma Chemicals. These are produced from isolates through chemical reactions, for example, l-carvone from limonene, cedryl acetate from cedrol, hydroxy citronellal from citronellal, or terpineol from pinene.

Synthetic Aroma Chemicals. These are synthesized from basic organic chemicals (from coal or petroleum). These are predominantly nature-identical aroma chemicals, that is, synthetic chemicals modeled after naturals, with the identical chemical structure, or, on a lesser scale, chemicals not found in Nature, new structures that resemble natural odor types nevertheless.

Arranging aroma chemicals by their chemical structures, gives the following general groupings: *Terpenes.* They are considered a special group due to their importance and close relationship, although strictly speaking, they belong to the aliphatic or alicyclic aroma chemicals. Examples: menthol, geraniol, citral. (Fig. 1)

Aromatics. Also known as benzenoids, the most important of these aroma chemicals are produced from benzene, toluene or phenol. They are the largest group, in terms of production as well as revenue.

Aliphatics. Unlike the terpenes and the aromatics, aliphatic compounds cannot be derived from just a few base compounds. Several alcohols, the aldehydes and some carbonic acid esters are of interest.

The saturated primary alcohols, produced by hydrogenation of corresponding carbonic acid esters (or other methods) have no value as odorants but are important starting points for the related aldehydes and esters.

However, one or two unsaturated alcohols, such as Hexene-3-ol-1 with its powerful "green" odor, are of interest to the perfumer.

Aliphatic aldehydes, especially the straight-chain saturated compounds with 8–13 carbon atoms but also some branched-chain and unsaturated compounds are of prime importance as perfume components. These aldehydes are produced either by dehydration of the corresponding alcohols, or from olefines by the oxo-process.

Amongst the aliphatic esters, which have a predominantly fruity odor, the acetic acid esters are the most widely used. Because of their tremendous strenght and their highly assertive odor-characteristics most of the aliphatic perfume components are used only in very small dosages.

Alicyclics (Cyclo-aliphatics). There are several very important fragrance materials amongst the non-terpenoid cyclo-aliphatic compounds. Prominent among them are the ketones, including cyclo-pentanone derivatives such as the important jasmin-components cis-jasmone and dihydro-methyl jasmonate, but also macro-cyclic ketones with 15 to 17 carbon atoms in the ring, such as muscone and civettone.

Although various syntheses of these macro-cyclic compounds have been worked out none of them have been used, so far, on an industrial scale, either because of poor yields or because the process entails too many reaction-stages.

Hetero-cyclic Compounds. Among the oxygen-heterocyclics, the lactones are of special significance. Like macro-cyclic ketones, the macro-cyclic esters are of outstanding odor value as musk odorants but are more easily produced by depolymerisation of corresponding linear polyesters; cyclo-pentadecanolide and ethylene brassylate are well-known examples.

Gamma-lactones, such as the gamma-undecalactone can be made by the single step process of radical addition of primary alcohols to acrylic acid.

Nitrogen and sulphur-heterocyclics (Pyridine, Pyrazine, Thiophene, Thiozole, etc.) are counted amongst the most odor-intensive trace components and used only in minute quantities.

Industrial Production of Synthetic Aroma Chemicals

Using terpene-products and aromatics as examples, the industrial production of important odorants from relatively few basic raw materials is illustrated by graphs, or "family trees"–see illustrations 1–7.

Terpene Compounds. (Fig. 1) For a long time, essential oils were the only available source of terpene aroma chemicals, which accordingly were isolates or semi-synthetic aroma chemicals. Ionones and methyl ionones from citral (ex-lemongrass or litsea cubeba oils) or hydroxycitronellal and l-menthol from d-citronellal (ex-citronella oil) are well known examples of these partial syntheses.

Since the 1960's, the natural sources could no longer cover the increasing world-wide demand for aroma chemicals for use in cosmetics and household products. At the same time, due to the synthetic production of vitamins A and E, an additional demand arose for citral and linalool, which hitherto were only available from essential oils.

The large vitamin producers solved the raw materials problem through the total synthesis of the terpene key ingredients from simple organic chemicals (acetone, acetylene, isoprene). The important intermediate products are methyl heptenone and dehydro linalool.

Partial syntheses are possible, starting from alpha or beta pinene obtained in abundant quantities by fractional distillation of turpentine oil (the low-boiling fraction of gum turpentine from conifers).

Myrcene is obtained from beta pinene by pyrolysis, and is converted by various processes to linalool or geraniol/nerol. The subsequent manufacture of products like citronellol, citral, ionone, etc., is carried out in the same manner as when starting materials were natural isolates.

For a long time, alpha pinene served in the manufacture of terpineol and isobornyl acetate, but only recently the industrial syntheses of linalool and geraniol/nerol from alpha pinene or alpha/beta pinene mixtures have been achieved. The process runs via cis-pinane, pinane hydroperoxide and pinanol to linalool, which ultimately is isomerized (rearranged) to geraniol/nerol.

Aromatics (Figs. 2–7)

All aromatics have the benzene ring in common and could be shown as one family. However, this would be too confusing, and would not do justice to certain raw materials and structural groupings.

The necessary raw materials are either directly obtained from coal or petroleum (benzene, toluene, xylene, naphtalene) or through chemically treating benzene or toluene (phenol, cresol).

From benzaldehyde, the main intermediate derivative of *toluene,* a large number of important aroma chemicals are synthesized, for example, cinnamic aldehyde, alpha amyl and alpha hexyl cinnamic aldehyde. Alkylation of toluene with isobutylene produces para tertiary butyl toluene, the starting material for lilial.

Phenyl ethyl alcohol is produced from *benzene,* and via cumol as an intermediate product, cyclamen aldehyde and galaxolide are produced.

Among the *cresols,* meta cresol has special significance for the syntheses of menthol (via thymol) and musk ambrette.

The *phenolics* represent the largest group of aroma chemicals, both tonnage-wise and in terms of revenue. Important aroma chemicals derived from phenol include: coumarin via salicylic aldehyde, anethol via anisole, and salicylates from salicylic acid. Pyro-catechol, that can also be obtained from phenol is, among others, the basic raw material for synthetic vanillin, isoeugenol and sandal.

Figures 1 through 7 demonstrate products that represent money-wise approximately 80 % of all synthetic aroma chemicals. Their industrial significance will increase in the coming decades. This positive outlook results from the fact that synthetically produced aroma chemicals–compared to natural–are available in consistent qualities, at relatively constant prices, and in practically any quantity desired.

Fragrances derived from terpenes

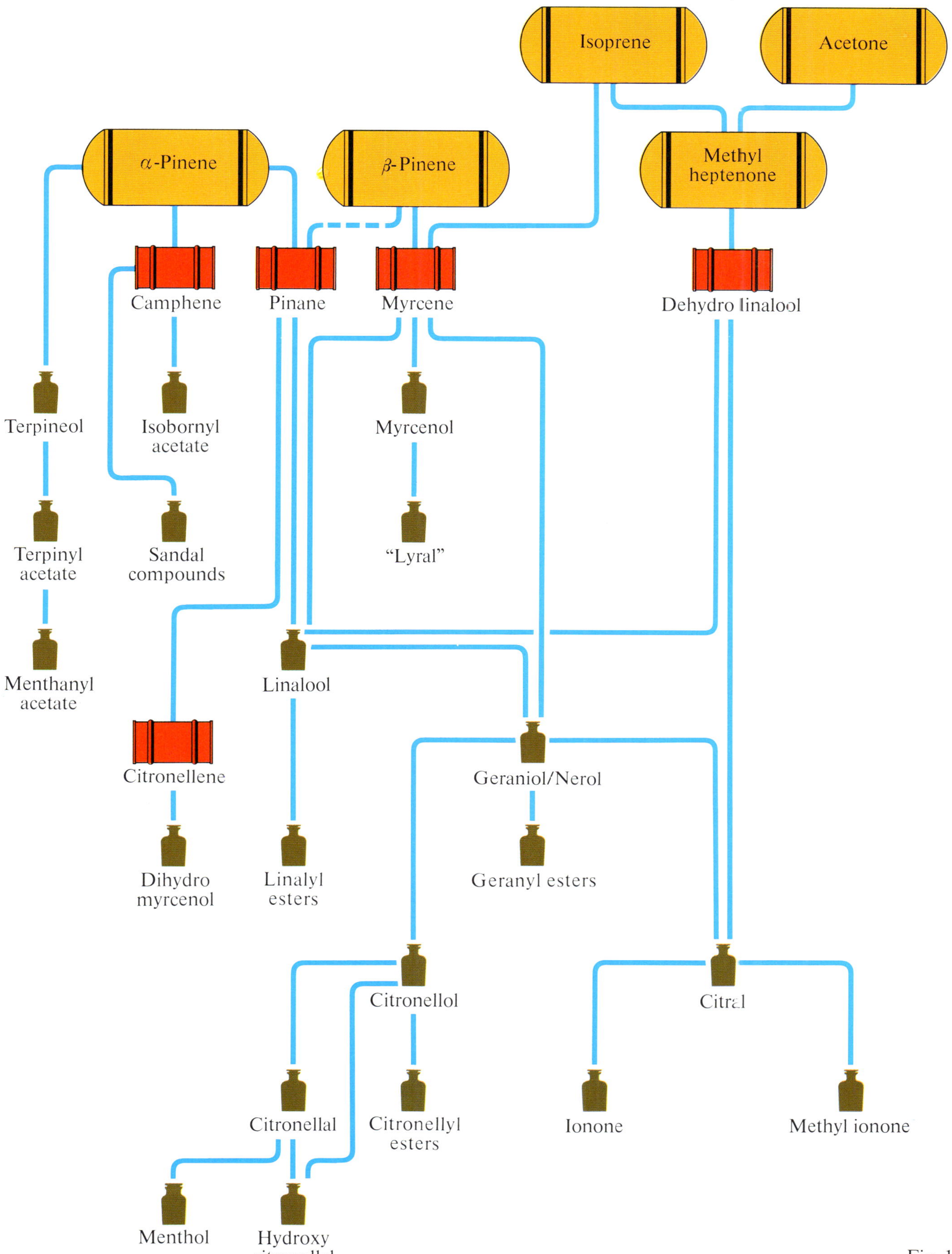

Fig. 1

Fragrances derived from toluene

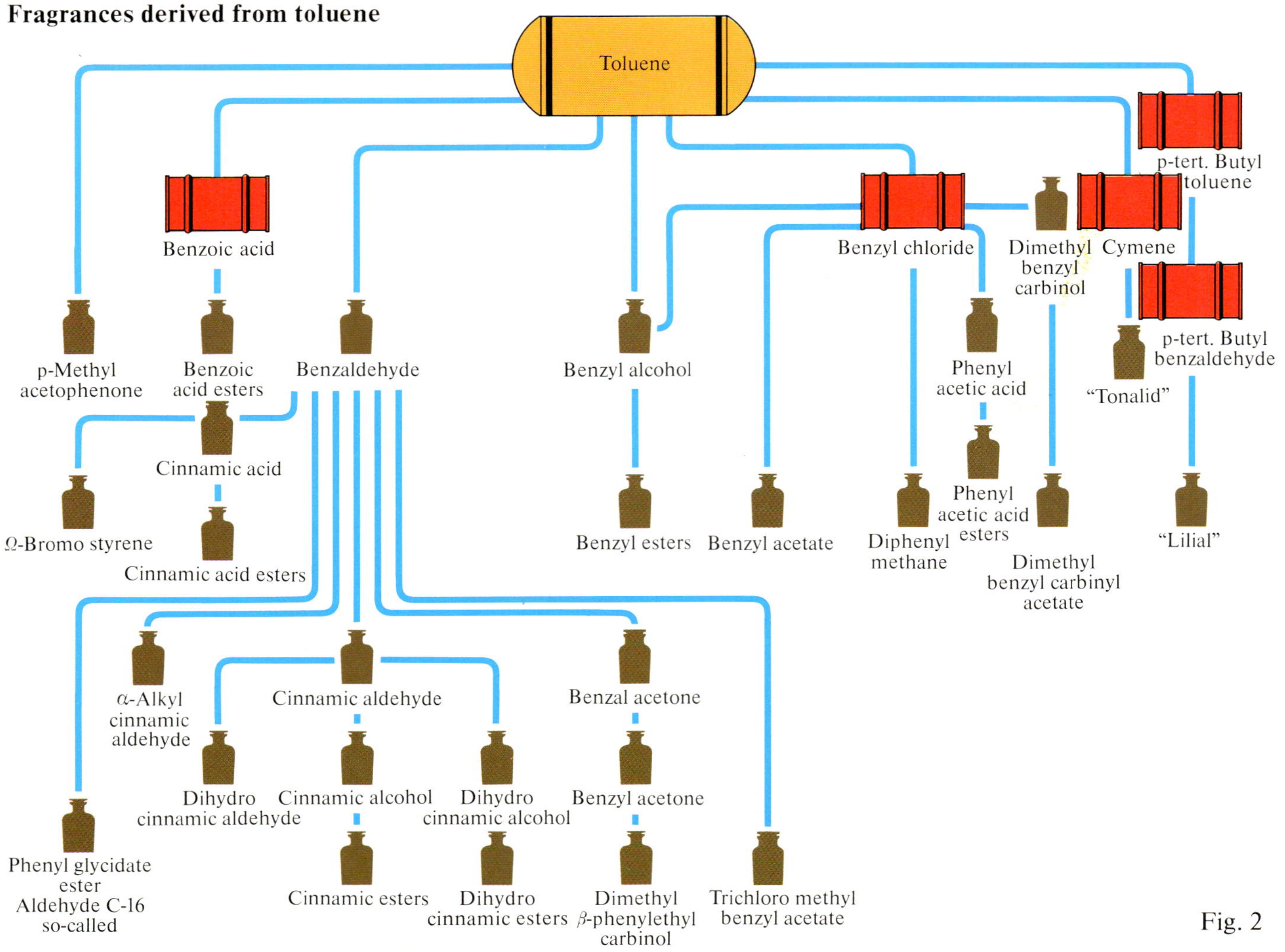

Toluene
Benzoic acid
Benzyl chloride
p-tert. Butyl toluene
Dimethyl benzyl carbinol
Cymene
p-tert. Butyl benzaldehyde
p-Methyl acetophenone
Benzoic acid esters
Benzaldehyde
Benzyl alcohol
Phenyl acetic acid
"Tonalid"
Cinnamic acid
Ω-Bromo styrene
Cinnamic acid esters
Benzyl esters
Benzyl acetate
Diphenyl methane
Phenyl acetic acid esters
Dimethyl benzyl carbinyl acetate
"Lilial"
α-Alkyl cinnamic aldehyde
Cinnamic aldehyde
Benzal acetone
Dihydro cinnamic aldehyde
Cinnamic alcohol
Dihydro cinnamic alcohol
Benzyl acetone
Phenyl glycidate ester
Aldehyde C-16
so-called
Cinnamic esters
Dihydro cinnamic esters
Dimethyl β-phenylethyl carbinol
Trichloro methyl benzyl acetate

Fig. 2

Fragrances derived from Naphthalene

Naphthalene
β-Naphthol
Phthalic acid
Decalol
Anthranilic acid
Decalyl ester
β-Naphthyl alkyl ether
Methyl β-naphthylketone
Methyl anthranilate
Phthalic acid esters
Methyl N-methylanthranilate acid methyl ester

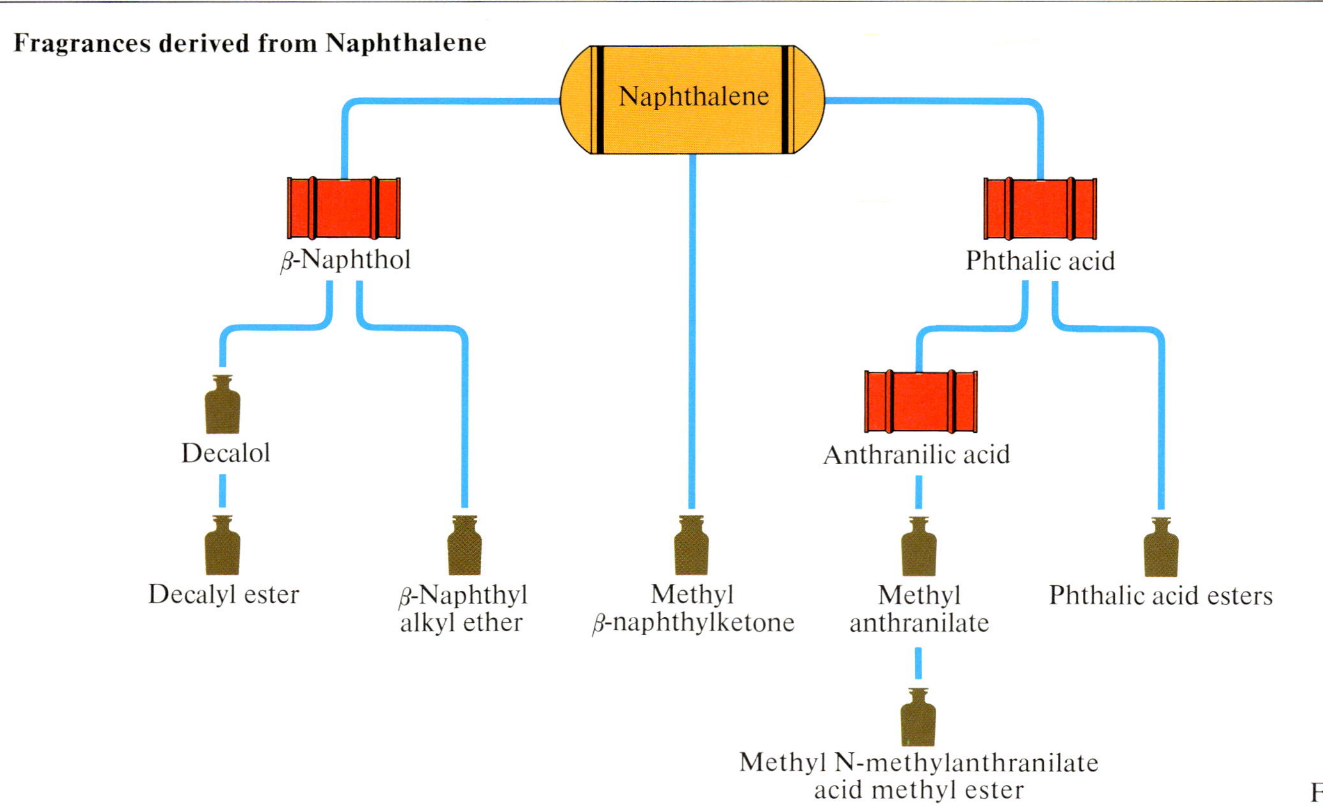

Fig. 3

Fragrances derived from benzene and m-xylene

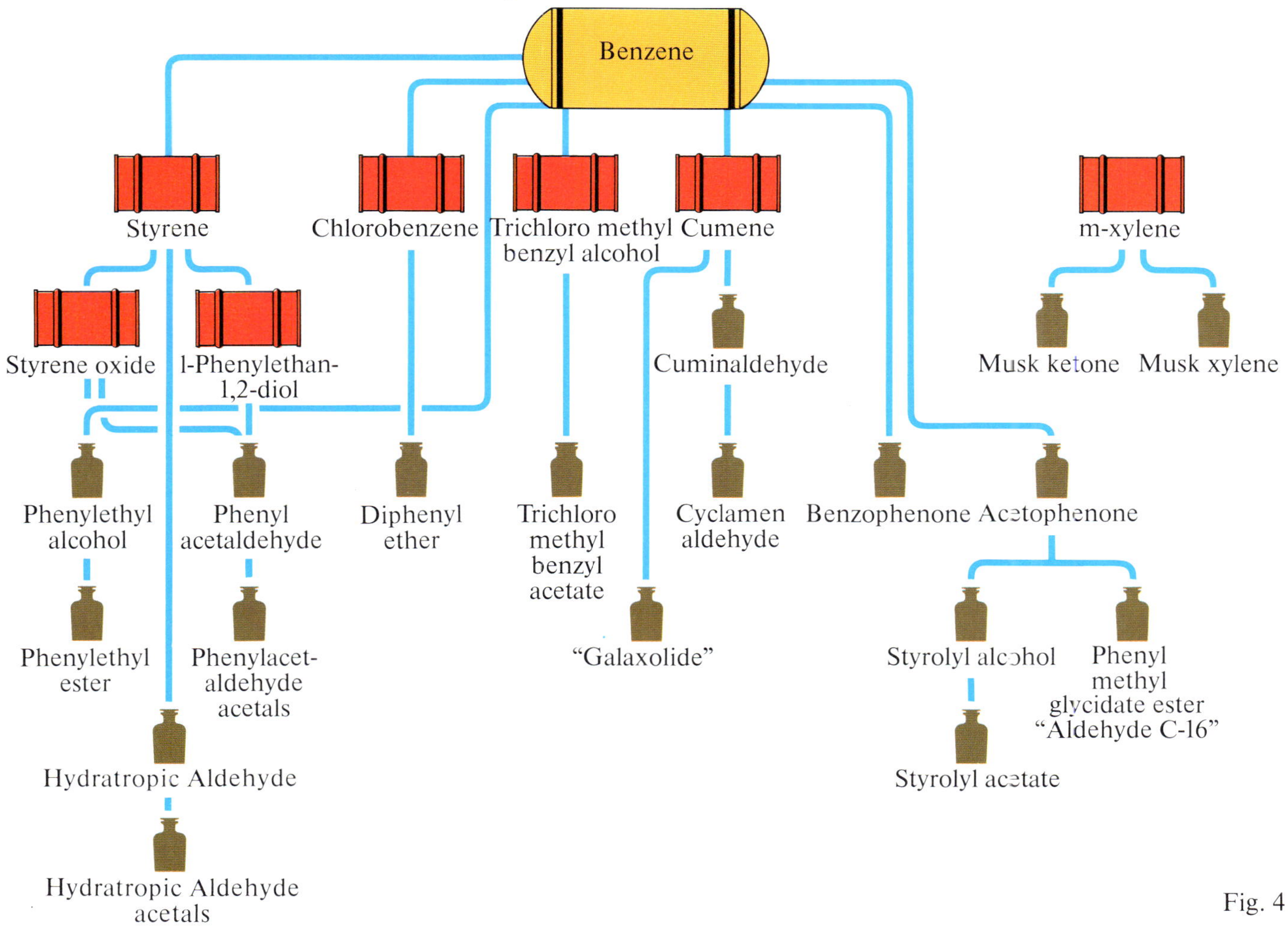

Benzene
Styrene
Chlorobenzene
Trichloro methyl benzyl alcohol
Cumene
m-xylene
Styrene oxide
1-Phenylethan-1,2-diol
Cuminaldehyde
Musk ketone
Musk xylene
Phenylethyl alcohol
Phenyl acetaldehyde
Diphenyl ether
Trichloro methyl benzyl acetate
Cyclamen aldehyde
Benzophenone
Acetophenone
Phenylethyl ester
Phenylacet-aldehyde acetals
"Galaxolide"
Styrolyl alcohol
Phenyl methyl glycidate ester "Aldehyde C-16"
Styrolyl acetate
Hydratropic Aldehyde
Hydratropic Aldehyde acetals

Fig. 4

Fragrances derived from the cresols

o-Cresol
m-Cresol
p-Cresol
p-Cresyl methyl ether
Coumarin
Thymol
Musk ambrette
Anisaldehyde
Dihydro cumarin
Menthol
Menthone
Anisalcohol
Menthyl ester
Anisyl esters

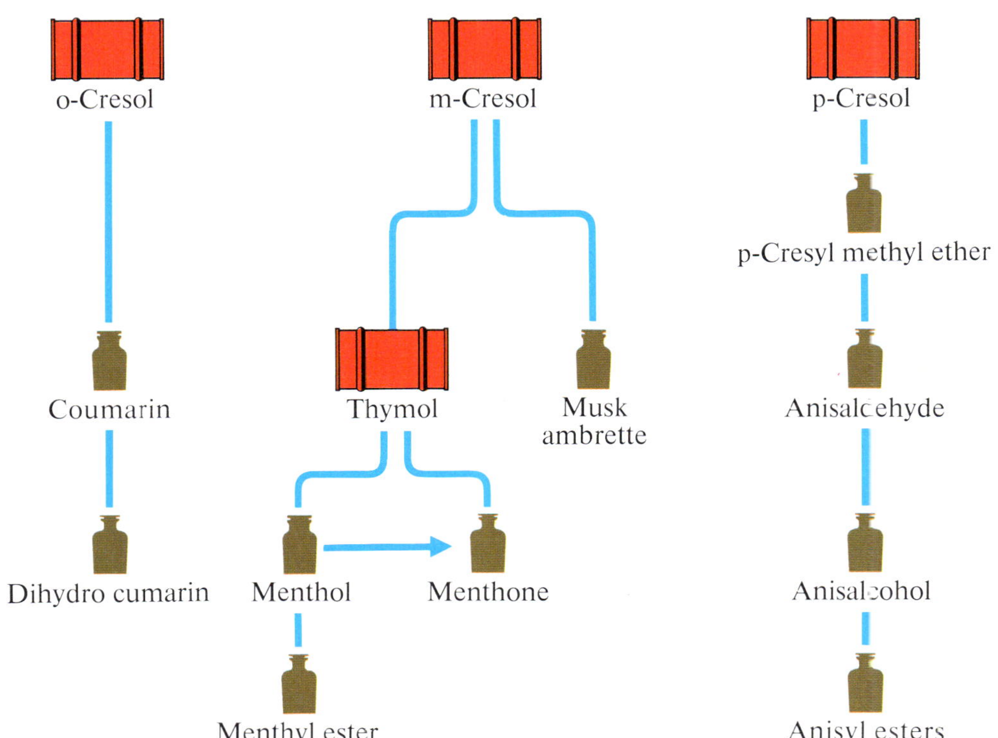

Fig. 5

Fragrances derived from phenol but not including pyrocatechol, resorcinol and hydroquinone derivatives.

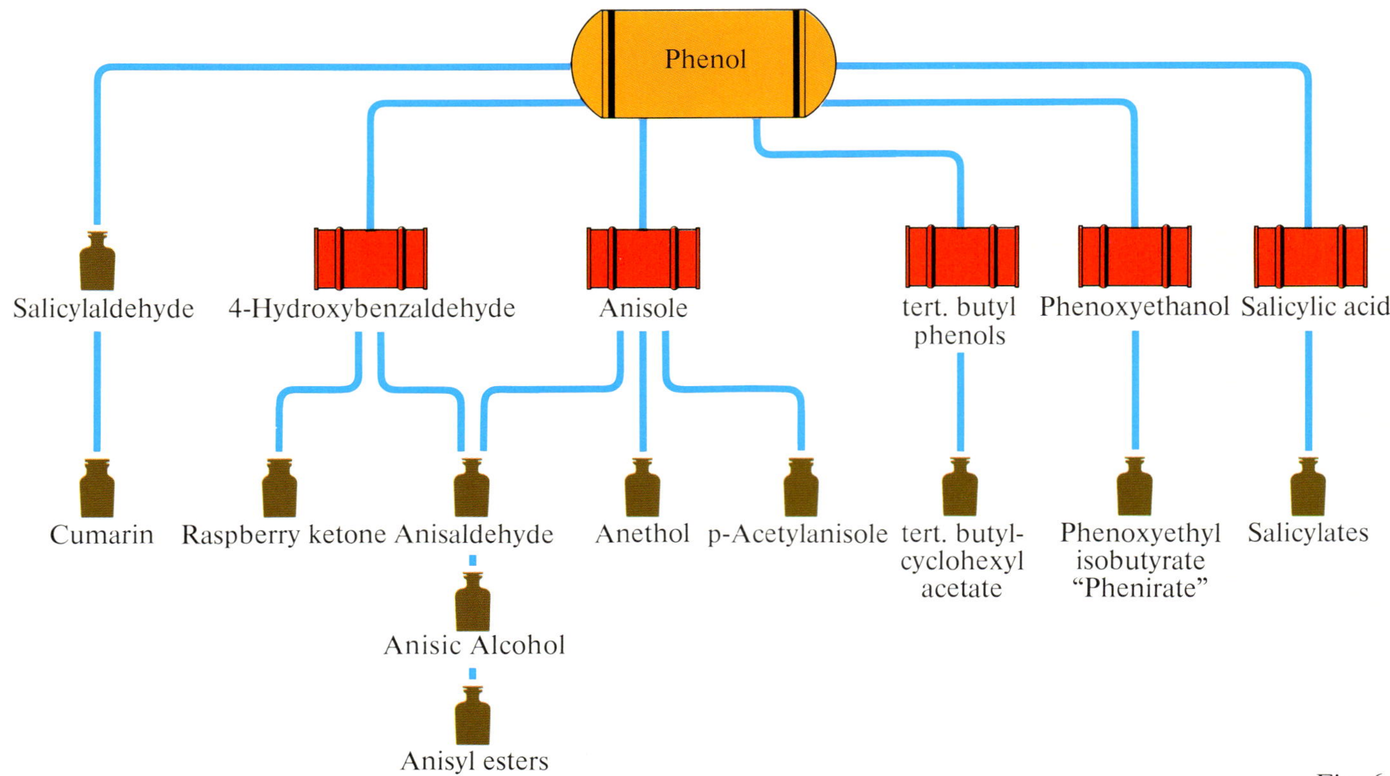

Fig. 6

Fragrances derived from pyrocatechol, resorcinol and hydroquinone

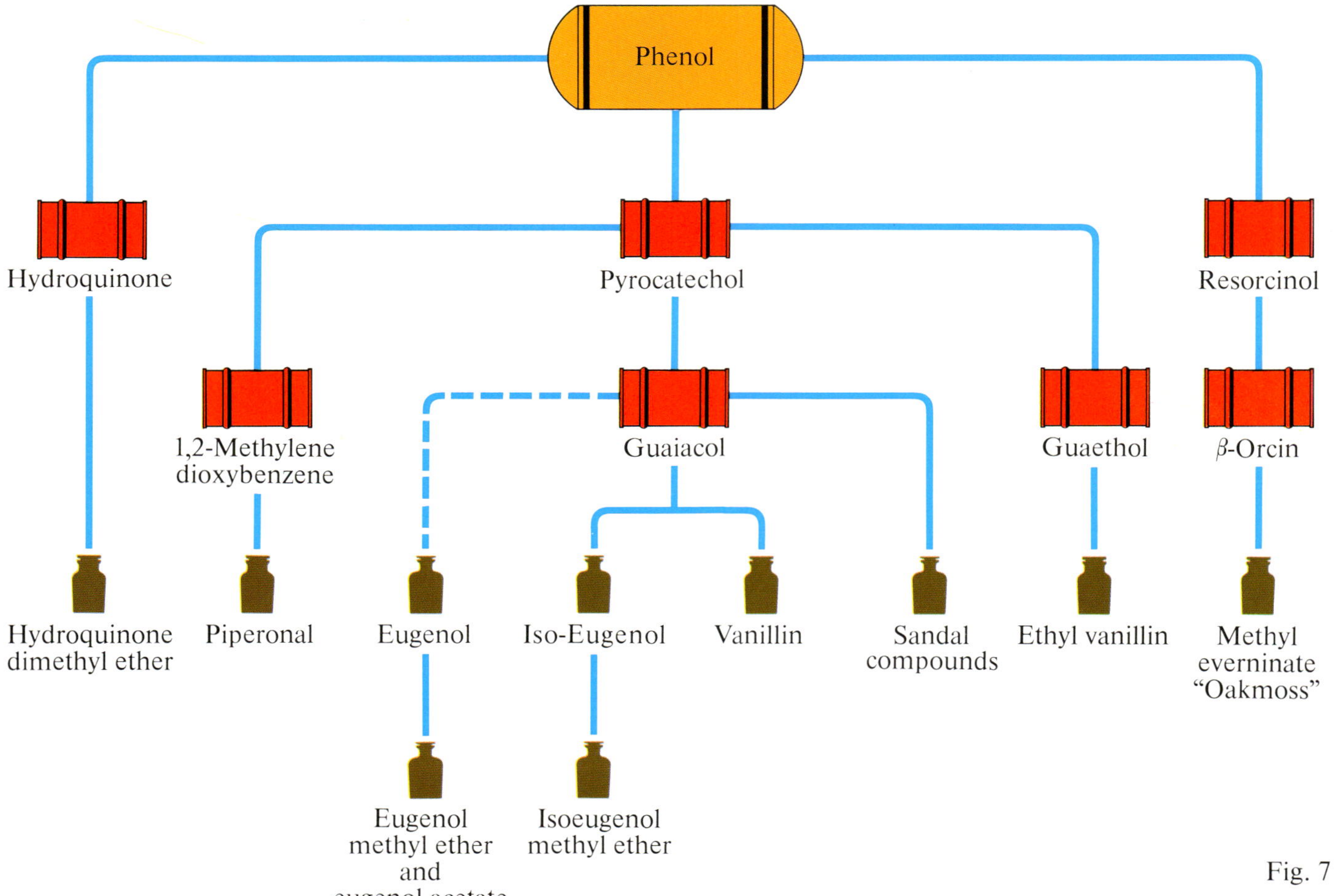

Fig. 7

Natural and Synthetic Fragrance Components

Defined according to: Source, Isolation, Appearance, Yield, Odor, Uses

H H
C6H5—C—C—OH
H H

A

Name **Acetanisole (p-Methoxy acetophenone) $C_9H_{10}O_2$**

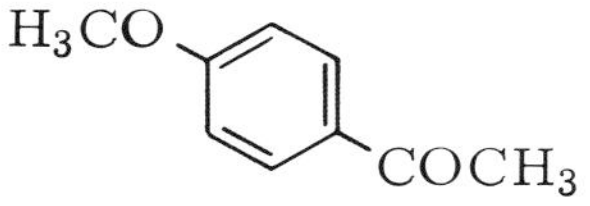

Source Ketone. Manufactured from anisole and acetyl chloride in the presence of aluminum chloride (Friedel-Crafts reaction). Found in anise and Piper longum.

Appearance White crystals.

Odor Narcotic, floral odor with warm, heavy and sweet undertones. Typical for lilac blossom.

Name **Acetyl cedrene $C_{17}H_{26}O$**

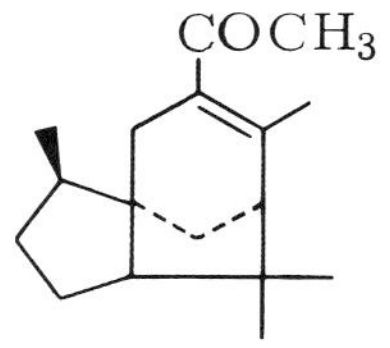

Source Ketone. Obtained via the acetylation of terpenes from cedar wood oil.

Appearance Light yellow, slightly viscous liquid.

Odor Harmonious, warm, woody odor reminiscent of vetiver and cedar.

Special Comments Acetyl cedrene appears on the market under various brand names.

Name **Acetophenone C_8H_8O**

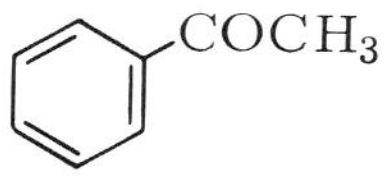

Source Ketone. Can be synthesized via the oxidation of ethyl benzene, but is also a by-product in the phenol synthesis. Occurs naturally in labdanum and cistus oils.

Appearance Colorless liquid.

Odor Sweet, slightly pungent odor, somewhat reminiscent of hawthorn, but with floral, coumarin-like undertones.

Name **6-Acetyl-1,1,2,4,4,6-hexamethyl tetrahydronaphthalene $C_{18}H_{26}O$**

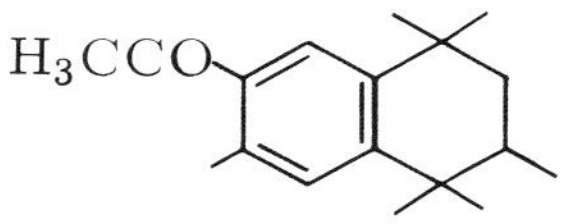

Source Synthetic tetralin-type musk. Is synthesized from para-cymene and dimethyl butene with subsequent acetylation.

Appearance Colorless to white crystalline material.

Odor Sweet, woody-musky odor.

Name **Alcohol C 10 (n-Decyl alcohol)**
$C_{10}H_{22}O$

$H_3C(CH_2)_8CH_2OH$

Source Alcohol. Can be obtained either from coconut fat via the hydrogenation of the corresponding fatty acid methyl ester or through oxidation and hydrolysis of the trialkyl aluminum derivatives.

Appearance Colorless liquid.

Odor Waxy and floral, mainly rosy odor with fatty-green undertones.

Name **Alcohol C 11 (Undecyl alcohol)**
$C_{11}H_{24}O$

$H_3C(CH_2)_9CH_2OH$

Source Alcohol. Can be synthesized via the catalytic hydrogenation of methyl undecylenate.

Appearance Colorless liquid.

Odor Slightly fatty, floral-fruity odor.

Name **Alcohol C 12 (Dodecyl alcohol)**
$C_{12}H_{26}O$

$H_3C(CH_2)_{10}CH_2OH$

Source Alcohol. Is synthesized in exactly the same way as alcohol C 10.

Appearance Colorless, oily liquid or crystalline mass.

Odor Very mild, slightly waxy, floral odor.

Name **Aldehyde C 7 (n-Heptanal)**
$C_7H_{14}O$

$H_3C(CH_2)_5CHO$

Source Aldehyde. Is obtained by the pyrolysis of castor oil.

Appearance Colorless liquid.

Odor Powerful, fatty, rancid odor displaying green, spicy undertones.

Special Comments It is an important raw material for the synthesis of other fragrance materials such as alpha-amyl cinnamic aldehyde and heptyl cyclopentanone.

Name **Aldehyde C 8 (n-Octanal)**
$C_8H_{16}O$

$H_3C(CH_2)_6CHO$

Source Aldehyde. Is obtained via the dehydrogenation of n-octanol. Occurs naturally in orange and rose oils.

Appearance Colorless liquid.

Odor Very powerful, fatty odor with distinct orange undertones.

Name **Aldehyde C 9 (n-Nonanal)**
$C_9H_{18}O$

$H_3C(CH_2)_7CHO$

Source Aldehyde. Can be obtained via the dehydrogenation of n-nonanol. Occurs naturally in a number of oils including citrus, rose, cinnamon and orris root.

Appearance Colorless liquid.

Odor Very powerful, wax-like odor with floral undertones.

Name	**Aldehyde C 10 (n-Decanal) $C_{10}H_{20}O$**
	$H_3C(CH_2)_8CHO$
Source	Aldehyde. Is synthesized via the dehydrogenation of n-decanol. Occurs naturally in citrus oils.
Appearance	Colorless liquid.
Odor	Powerful, somewhat sweet, citrus peel-like odor with slightly rancid-fatty notes.

Name	**Aldehyde C 11 (n-Undecanal) $C_{11}H_{22}O$**
	$H_3C(CH_2)_9CHO$
Source	Aldehyde. Is manufactured via the dehydrogenation of n-undecanol. Occurs naturally in citrus oils.
Appearance	Colorless liquid.
Odor	Very powerful, waxy, fatty odor with floral undertones.

Name	**Aldehyde C 11 (10-Undecen-1-al) $C_{11}H_{20}O$**
	$H_2C{=}CH(CH_2)_8CHO$
Source	Aldehyde. Is synthesized via the Rosenmund reduction of the acid chloride of undecylenic acid.
Appearance	Colorless liquid.
Odor	Powerful waxy-rose odor with metallic, green undertones.

Name	**Aldehyde C 12 (Lauric) (n-Dodecanal) $C_{12}H_{24}O$**
	$H_3C(CH_2)_{10}CHO$
Source	Aldehyde. Is obtained via the dehydrogenation of lauryl alcohol. Occurs naturally in pine needle oil.
Appearance	Colorless liquid.
Odor	Fresh, piney, herbaceous, floral odor with a slight wax note.

Name	**Aldehyde C 12 MNA (Methyl nonyl acetaldehyde) $C_{12}H_{24}O$**
	$H_3C(CH_2)_8CH(CH_3)CHO$
Source	Aldehyde. Is synthesized from methyl nonyl ketone and ethyl monochloroacetate via the glycidic ester.
Appearance	Colorless liquid.
Odor	Dry, slightly fruity odor reminiscent of ambergris and incense and with floral, waxy notes.

Name **Ambergris tincture**

Source: Ambergris is a greyish-black pathological excretion from the sperm whale, Physeter macrocephalus, and is often found washed up on beaches. It is also a valuable by-product of the whaling industry.

Isolation Procedure: Via extraction of the powdered ambergris.

Odor: The odor can vary and exhibits different nuances such as woody, dry balsamic, somewhat tobacco-like notes and also has an erogenic note.

Uses: Mainly in expensive perfume oils. However, nowadays one usually uses substitutes because genuine ambergris is extremely difficult to obtain.

Sperm whale – Physeter macrocephalus

Name **Ambrette seed absolute**

Botanical Source/ Occurrence: From ambrette seeds which are produced in the fruits of a cultivated plant, Hibiscus abelmoschus. The plant belongs to the mallow family and is to be found in Central and South America, Indonesia and India.

Isolation Procedure: The dried and powdered seeds are subject to steam distillation whereby an essential oil is obtained which is very rich in fatty acids. These may be removed by solvent extraction, thus yielding the absolute.

Yield: 0.3 – 0.5 % depending upon the source of the plant material.

Odor: Floral, musk-like, slightly sweet odor with a distinct Cognac note.

Uses: Finds use in expensive fine fragrances.

Name **alpha-Amyl cinnamic aldehyde** **$C_{14}H_{18}O$**

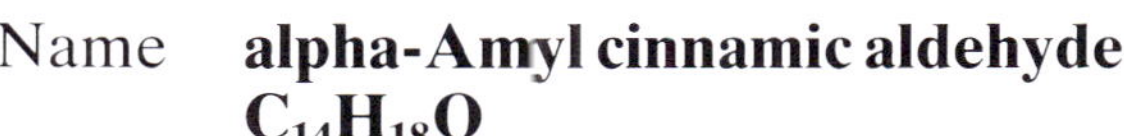

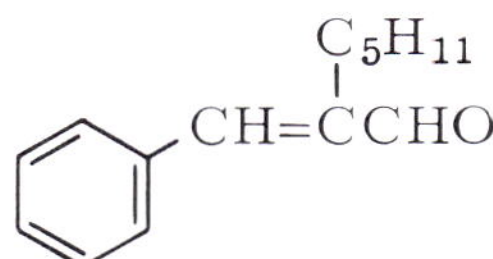

Source: Aldehyde. Can be synthesized via the condensation of benzaldehyde and n-heptanal.

Appearance: Yellow liquid.

Odor: Intense floral, jasmin-like odor with a warm dry-out.

Name **Anethole (iso-Estragole) $C_{10}H_{12}O$**

H_3CO–C_6H_4–$CH{=}CHCH_3$

Source Phenol ether. Can be isolated from star anise oil or from turpentine oil. Is obtained synthetically from anisole and propionaldehyde. Occurs naturally in anise, star anise and fennel oils.

Appearance White crystalline substance, but can also be a colorless liquid.

Odor Very sweet, herbaceous-warm typical odor of star anise.

Name **Angelica root oil**

Botanical Source/ Occurrence This oil is obtained from the roots of the plant, Angelica archangelica, which is an umbellifer and is cultivated in Central Europe.

Isolation Procedure The oil is steam distilled from the fresh or dried comminuted roots.

Yield 0.3–0.4 %.

Odor Exhibits an earthy, somewhat musk-like, peppery, aromatic odor with a green, spicy top note.

Uses Used extensively as a toner in masculine notes.

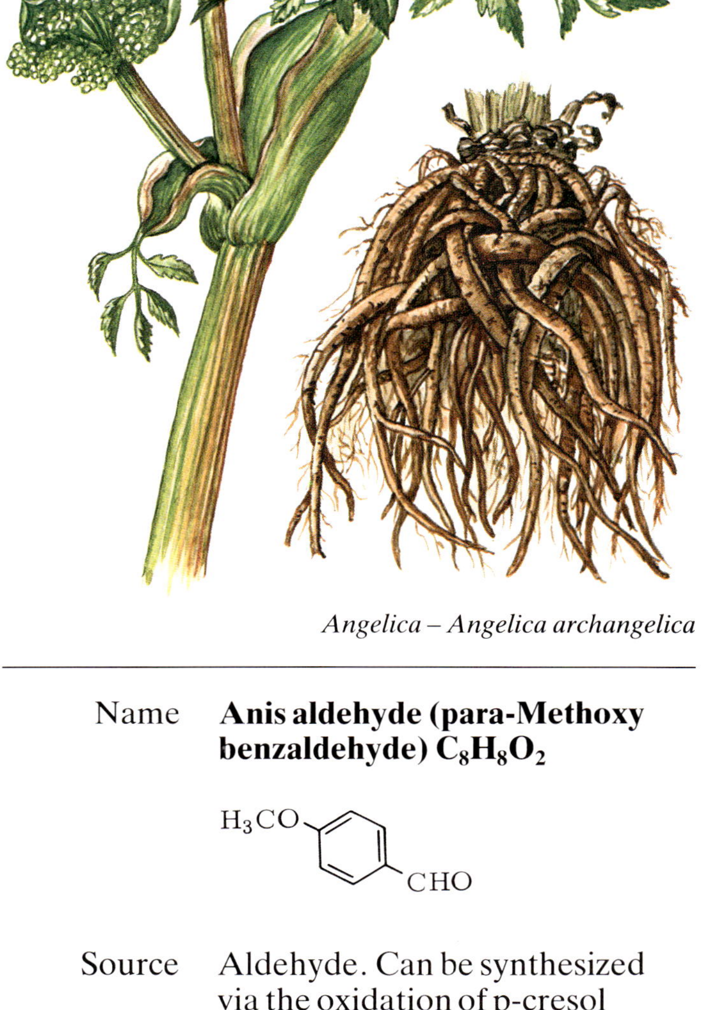

Angelica – Angelica archangelica

Name **Anis aldehyde (para-Methoxy benzaldehyde) $C_8H_8O_2$**

H_3CO–C_6H_4–CHO

Source Aldehyde. Can be synthesized via the oxidation of p-cresol methylether. Occurs naturally in vanilla, anise and fennel oils.

Appearance Pale yellow liquid.

Odor Powerful, sweet, hay-like odor with slight cresol notes.

Name	**Anise oil / Star anise oil**
Botanical Source/ Occurrence	From the seeds of the tree, Illicium verum, which belongs to the magnolia family and is cultivated mainly in China, Spain and the U.S.S.R. Anise oil is obtained from the plant Pimpinella anisum.
Isolation Procedure	The comminuted seeds are steam distilled.
Yield	2–3 % depending on the raw material.
Odor	Very powerful, typical odor which is sweet, herbaceous and lively.
Uses	As a toner in many different perfume types.

Name	**Artemisia oil**
Botanical Source/ Occurrence	Obtained from the composite plants, Artemisia vulgaris and Artemisia alba, which are cultivated in Morocco, Algeria, Yugoslavia and France.
Isolation Procedure	The essential oil is steam distilled from the dried herb.
Yield	Approx. 0.3 %.
Odor	Very bright, lively, somewhat herbaceous, spicy odor.
Uses	Used widely in many types of nature-based perfumes and also in balsamic complexes.

Anise – Pimpinella anisum

Artemisia – Artemisia vulgaris

Name	**Asafoetida resinoid**
Botanical Source/ Occurrence	From the dried, milky exudate of Ferula asa foetida which is an umbellifer plant native to Iran and Afghanistan.
Isolation Procedure	Via extraction of the dried plant resinous gum.
Yield	Varies depending upon the raw material and extraction procedure.
Odor	Very intense onion-, garlic-like odor.
Uses	Is used as a toner, but is somewhat limited due to it intensity.

B

Name	**Basil oil**
Botanical Source/ Occurrence	Is obtained from the small plant Ocimum basilicum, which is cultivated in Yugoslavia, France, Italy, the U.S.A., Eastern Europe, Madagascar, Réunion and the Seychelles.
Isolation Procedure	The essential oil is steam distilled from the dried herb.
Yield	0.1–0.2 %.
Odor	Very bright, lively, somewhat herbaceous, spicy odor.
Uses	Used widely in many types of nature-based perfumes and also in balsamic complexes.

Asafoetida – Ferula asa foetida

Basil – Ocimum basilicum

Name	**Bay oil**
Botanical Source/ Occurrence	Is obtained from the leaves of the bay tree, Pimenta racemosa, which is found in the West Indies, South and Central America.
Isolation Procedure	The oil is obtained by steam distillation of the leaves.
Yield	0.5 – 1.5 %.
Odor	Very powerful, spicy, sweet odor with a distinct clove note.
Uses	Is a frequently used component of fresh and spicy, fantasy creations.

Name	**Beeswax absolute**
Botanical Source/ Occurrence	From the wax of the honey bee, Apis mellefica.
Isolation Procedure	Via extraction of the beeswax.
Yield	0.1 – 0.2 %.
Odor	Mild, oily, sweet, honey-like odor with herbaceous notes.
Uses	Finds use in floral fine fragrances.

Bay – Pimenta racemosa

Honeycomb

Name **Benzaldehyde C_7H_6O**

(structural formula: benzene ring with CHO)

Source Aldehyde. Can be synthesized from toluene via benzal chloride as intermediate. Occurs naturally in bitter almond oil.

Appearance Colorless liquid.

Odor Very powerful, sweet odor typical of bitter almonds.

Special Comments It is also sold as artificial bitter almond oil.

Benzoin Siam – Styrax tonkinensis

Name **Benzoin Siam resinoid**

Botanical Source/ Occurrence Is obtained from the resin of the tree, Styrax tonkinensis, which grows wild in Thailand, Laos, Cambodia and Vietnam.

Isolation Procedure Via solvent extraction of the resin.

Yield 85 – 95 %, depending on the source of the plant material.

Odor Sweet, balsamic, chocolate-like odor of great tenacity.

Uses For balsamic notes and as a fixative.

Name **Benzoin Sumatra resinoid**

Botanical Source/ Occurrence From the resin of the tree, Styrax benzoin, which grows wild mainly on Sumatra.

Isolation Procedure Via solvent extraction of the resin.

Yield 65 – 85 %.

Odor Displays a warm, sweet, powdery, balsamic odor.

Uses As a fixative.

Name **Benzophenone $C_{13}H_{10}O$**

$C_6H_5COC_6H_5$

Source Ketone. Made from benzoyl chloride and benzene.

Appearance White crystals.

Odor Consists of sweet, floral and dry elements mixed with slight medicinal and chemical notes.

Name **Benzyl acetate $C_9H_{10}O_2$**

$C_6H_5CH_2OCOCH_3$

Source Ester. Can be synthesized from benzyl alcohol and acetic anhydride. It occurs naturally in jasmin, gardenia and ylang-ylang oils.

Appearance Colorless liquid.

Odor Fruity, somewhat acidic, jasmin-like odor.

Name **Benzyl alcohol C_7H_8O**

$C_6H_5CH_2OH$

Source Alcohol. Is synthesized from toluene via benzyl chloride as intermediate. Occurs naturally in many essential oils, especially ylang-ylang, jasmin, tuberose and wallflower.

Appearance Colorless liquid.

Odor Very weak odor.

Name **Benzyl benzoate $C_{14}H_{12}O_2$**

$C_6H_5COOCH_2C_6H_5$

Source Ester. Synthesized from benzyl chloride and sodium benzoate. Occurs naturally in Tolu and Peru balsam and in tuberose oil.

Appearance Colorless liquid or white crystals.

Odor Faint, nondescript odor.

Name **Benzyl salicylate $C_{14}H_{12}O_3$**

$C_6H_4(OH)COOCH_2C_6H_5$

Source Ester. Can be synthesized from methyl salicylate and benzyl alcohol via transesterification.

Appearance Colorless liquid.

Odor Generally floral, azalea-type odor with undertones associated with hawthorn, tulip and daffodils.

Name	**Bergamot oil**
Botanical Source/ Occurrence	From the peels of the fruit of the bergamot tree, Citrus bergamia, which is native to Italy, South America and Western Africa.
Isolation Procedure	The oil is produced by cold expression from the peel of the nearly ripe but still green fruit. This used to be performed manually, but is now done by machines.
Yield	Approx. 0.5 %, depending upon the source and the production procedure.
Odor	Has a fresh, clear, lively odor, somewhat fruity and sweet which displays great orginality.
Uses	Due to its freshness this oil is used extensively, especially in Eau de Cologne and Eau de Toilette.

Name	**Birch tar oil**
Botanical Source/ Occurrence	The tar is produced from the various birch species, including Betula alba, which grow wild in Europe.
Isolation Procedure	Via destructive distillation of the wood followed by rectification.
Yield	Varies depending upon the raw material and production procedure.
Odor	Displays a woody, tarry, smoky odor with pleasantly sweet, oily leather-like notes.
Uses	Due to its very special nature, this powerful and highly appreciated material is used only in traces, mainly in masculine fragrances.

Bergamot – Citrus bergamia

Birch – Betula

Name	**Borneol $C_{10}H_{18}O$**
Source	Alcohol. Is synthesized from pinene via bornyl chloride as intermediate. It occurs naturally in rosemary and lavender oils.
Appearance	Colorless, crystalline material.
Odor	Powerful, pungent, dry-camphoraceous, woody-peppery odor, somewhat medicinal.

Name	**l-Bornyl acetate $C_{12}H_{20}O_2$**
Source	Ester. Can be obtained from borneol and acetic anhydride or from turpentine oil or pinene via the Wagner-Meerwein rearrangement. Occurs naturally in many conifer oils.
Appearance	Colorless liquid.
Odor	Powerful, herbaceous odor reminiscent of pine needles.

Name	**Boronia absolute**
Botanical Source/ Occurrence	The raw material for this product is the small bushy plant, Boronia megastigma, which grows wild in Western Australia.
Isolation Procedure	Via extraction of the flowers which first yields the concrète. Extraction of the concrète then gives rise to the absolute.
Yield	0.1–0.2 %.

Boronia – Boronia megastigma

Odor	Fruity, spicy-herbaceous odor with sweet, rose-like and floral undertones.
Uses	In high-class chypre and fougère perfumes.

Name	**omega-Bromostyrene C_8H_7Br**
Source	Halogenated hydrocarbon. Manufactured from cinnamic acid and bromine, whereby the intermediate dibromophenyl propionic acid is transformed to bromostyrene by the action of soda.
Appearance	Yellow liquid.
Odor	Very strong impact with a distinct chemical character. Exhibits a hyacinth note when diluted.

Name	**Broom absolute**
Botanical Source/ Occurrence	Is obtained from the blossoms of the so-called Spanish broom, Spartium junceum. The small shrub grows wild throughout the whole of the mediterranean area and is cultivated for production purposes in Southern France and Morocco.
Isolation Procedure	Broom absolute is produced from the concrète which in turn is obtained via solvent extraction of the blossoms.
Yield	Approx. 0.2 % concrète which gives some 20 – 40 % absolute.
Odor	Sweet, floral, hay-like fragrance which has bitter undertones and great orginality.
Uses	Frequently used in floral fine fragrances.

Broom – Spartium junceum

Name	**Bruyère absolute**
Botanical Source/ Occurrence	Is obtained from the roots of the heather shrub, Erica arborea, which grows wild throughout the whole of the mediterranean area.
Isolation Procedure	Via extraction of the comminuted roots.
Yield	0.08 – 0.1 %.
Odor	Mild, balsamic, spicy-herbaceous odor with dry, woody notes.
Uses	Is sometimes incorporated into chypre bases and masculine notes.

Bruyère – Erica arborea

Name	**Buchu leaf oil**
Botanical Source/ Occurrence	The oil is obtained from the herbs, Barosma betulina and Barosma crenulata, which grow wild and are also cultivated in South Africa.
Isolation Procedure	The fresh leaves are steam distilled.
Yield	0.8–2.5 %.
Odor	Very strong, minty-camphoraceous odor which is very reminiscent of black currants.
Uses	Is only used in very small amounts due to its intensity.

Buchu – Barosma Species

Name	**n-Butyl acetate $C_6H_{12}O_2$**
	$H_3CCOO(CH_2)_3CH_3$
Source	Ester. Made from n-butyl alcohol and either acetic acid or acetic anhydride.
Appearance	Colorless liquid.
Odor	Sweet and very volatile solvent-like odor; somewhat fruity and acidic. Nail polish remover.

Name	**n-Butyl butyrate $C_8H_{16}O_2$**
	$H_3CCH_2CH_2COO(CH_2)_3CH_3$
Source	Ester. Synthesized from n-butyl alcohol and n-butyric acid.
Appearance	Colorless liquid.
Odor	Fresh and sweet-fruity, reminiscent of pineapples and bananas, somewhat cheesy.

Name	**ortho-tertiary-Butyl cyclohexyl acetate $C_{12}H_{22}O_2$**
	Cyclohexane ring with $OCOCH_3$ and $C(CH_3)_3$
Source	Ester. Made by the hydrogenation of o-tert.-butyl phenol followed by acetylation.
Appearance	Colorless liquid.
Odor	Displays a floral, woody odor with a slightly acidic, warm dry-out and a distinctly fresh, fruity top note.
Special Comments	This product is usually a mixture of the cis- and trans-isomers.

Name **para-tertiary-Butyl-alpha-methyl dihydrocinnamic aldehyde $C_{14}H_{20}O$**

$$(H_3C)_3C-C_6H_4-CH_2CH(CH_3)CHO$$

Source Aldehyde. Is synthesized via the condensation of para-tert.-butylbenzaldehyde with propionaldehyde and subsequent catalytic hydrogenation.

Appearance Colorless liquid.

Odor Delicate floral odor, somewhat green, soft and reminiscent of tulip, lily and lily of the valley.

Name **para-tertiary-Butyl cyclohexyl acetate $C_{12}H_{22}O_2$**

$$(H_3C)_3C-C_6H_{10}-OCOCH_3$$

Source Ester. Can be synthesized by the catalytic hydrogenation of para-tert.-butyl phenol followed by acetylation.

Appearance Colorless liquid.

Odor Delicate sweet-soft, woody odor with a pronounced fruity-floral note which makes it resemble orris, ionone and cedarwood.

C

Name **Cabreuva oil**

Botanical Source/ Occurrence From the wood of various species of wild-growing Myrocarpus trees, for example Myrocarpus frondosus and Myrocarpus fastigiatus, which are to be found in Brazil, Paraguay and Argentina.

Isolation Procedure Is produced by steam distillation of the waste wood, i.e. chippings and sawdust from the saw mills.

Yield 1.6–1.8%.

Odor Warm, woody, fatty odor.

Uses As a low cost woody fragrance component in fantasy creations.

Cade – Juniperus oxycedrus

Name	**Cade oil (Juniper tar oil)**
Botanical Source/ Occurrence	From the wood of the shrub, Juniperus oxycedrus, which grows mainly in Europe.
Isolation Procedure	Via destructive distillation of the comminuted wood.
Yield	1.4 – 1.6 %.
Odor	Intense, smoky, leathery, tar-like odor.
Uses	Lends strength and originality when used in small quantities in leather bases, fougères or pine compositions.

Name	**Cajuput oil**
Botanical Source/ Occurrence	From the leaves and twigs of a medium-sized tree, Melaleuca minor and possibly other species of Melaleuca, which grow wild in Indonesia, Australia and Malaysia.
Isolation Procedure	Via steam distillation of the fresh leaves and twigs.
Yield	0.8 – 1 %, depending upon the source of the plant material and the production procedure.
Odor	Very powerful, herbaceous, eucalyptus-like odor.
Uses	The oil is difficult to come by and is therefore not used much.

Name	**Calamus oil**
Botanical Source/ Occurrence	From the plant, Acorus calamus, which grows wild but is also cultivated in Korea, India and the U.S.S.R.
Isolation Procedure	Via steam distillation of the fresh or dried roots.
Yield	Between 1.2 and 1.5 % depending upon the source of the raw material.
Odor	Heavy, earthy, slightly sweet and aromatic root-like odor with bitter undertones.
Uses	As a toner in spicy, herbal perfumes.

Calamus – Acorus calamus

Name	**Calendula concrète / Calendula absolute**
Botanical Source/ Occurrence	From the flowers of the marigold Calendula officinalis, which belongs to the Compositae family. The plant is cultivated in Europe and North America.
Isolation Procedure	Extraction of the flowers produces the concrète from which the absolute is obtained.
Yield	0.3 – 0.4 % as concrète from which some 30 % absolute is obtained.
Odor	Typical, very intense, bitter, herbaceous odor.
Uses	This oil is rarely used due to the extremely inconsistant supply situation.

Marigold – Calendula officinalis

Name	**Camphor $C_{10}H_{16}O$**

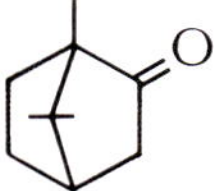

Source	Ketone. Natural camphor is obtained from the essential oil of the camphor tree (Cinnamomum camphora) and the synthetic material from pinene.
Appearance	Crystalline, transparent mass.
Odor	Exhibits a very intense, distinctive, typically medicinal odor.

Name	**Cananga oil**
Botanical Source/ Occurrence	From the flowers of the tree, Cananga odorata, which grows wild, but is also cultivated on Madagascar, Java, the Philippines and the Comoro Islands.
Isolation Procedure	Via steam distillation of the flowers.
Yield	0.8 – 1.0 %.
Odor	The oil displays a warm, sweet, floral and narcotic odor.
Uses	Is used in floral fragrances to ellicit both a sweet and secretive but also a lively, alert response.

Name	**Caraway oil**
Botanical Source/ Occurrence	From the seeds of the caraway plant, Carum carvi, a small umbellifer herb which is cultivated throughout Europe, but especially in Yugoslavia and Poland.
Isolation Procedure	This oil is steam distilled from the dried, crushed, ripe seeds.
Yield	3–5 % and even more from good seeds.
Odor	Displays a very typical, intensely spicy, caraway odor.
Uses	Due to its extreme intensity, this oil is only used in small amounts in modern masculine fragrances.

Name	**Cardamom oil**
Botanical Source/ Occurrence	The oil is obtained from Elettaria cardamomum Maton, a plant belonging to the ginger family. This plant grows wild and is also cultivated in India, Tanzania and Guatemala.
Isolation Procedure	Via steam distillation of the seeds.
Yield	1–4.5 %.
Odor	Spicy, aromatic odor with balsamic, floral undertones.
Uses	Finds use as a spicy toner in both masculine and feminine perfumes.

Caraway – Carum carvi

Cardamom – Elettaria cardamomum Maton

Name	**Carrot seed oil**
Botanical Source/ Occurrence	Is obtained from the seeds of the common carrot, Daucus carota, which belongs to the umbellifer family. Even though carrots are a popular vegetable worldwide, the oil is produced only in France, Hungary, Yugoslavia and U.S.S.R.
Isolation Procedure	This oil is steam distilled from the dried seeds.
Yield	Approx. 1.5 %.
Odor	Displays a fatty, earthy, woody, root-like odor of considerable diffusion.
Uses	In fougères and chypres and also in many natural-type fragrances.

Carrot – Daucus carota

Name	**Carvone $C_{10}H_{14}O$**

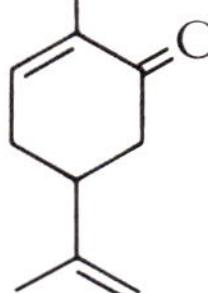

Source	Ketone. Exists in nature as two optically active isomers. Found in many essential oils from which it is obtained by distillation. l-Carvone is especially found in spearmint oil and d-carvone in dill and caraway seed oil. Can be synthesized from limonene such that l-carvone is obtained from d-limonene and d-carvone from l-limonene.
Appearance	Colorless, light yellow liquid.
Odor	Very attractive, typical odor of spearmint, caraway and dill only even cleaner, clearer and sweeter.

Name	**beta-Caryophyllene $C_{15}H_{24}$**

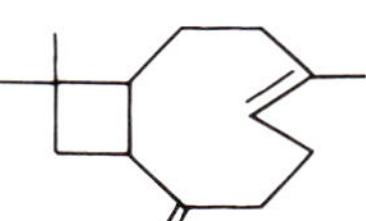

Source	Sesquiterpene hydrocarbon. Obtained by the distillation of clove leaf or clove stem oil. Found in numerous essential oils.
Appearance	Colorless liquid.
Odor	Somewhat spicy, acidic, woody odor. Dry and reminiscent of vetiver.

Name	**Cascarilla oil**
Botanical Source/ Occurrence	This oil is obtained from Croton eluteria Benett and Croton cascarilla Benett, which are small trees which grow wild in the West Indies.
Isolation Procedure	Via steam distillation of the dried bark.
Yield	1.2–3 %.
Odor	Intensely spicy-aromatic, slightly peppery and cinnamon-like odor, reminiscent of kitchen herbs.
Uses	The oil presents very interesting spicy and fresh notes in chypre bases and masculine fragrances when used at low levels.

Name	**Cassia oil (Chinese Cinnamon oil)**
Botanical Source/ Occurrence	From Cinnamomum cassia, a large slender evergreen tree, which grows mainly in China, but may also be found in South East Asia.
Isolation Procedure	Via steam distillation of the leaves, stalks and young twigs of the tree.
Yield	0.5–1.9 % depending upon the raw material.
Odor	Very intense, spicy, cinnamon-like odor.
Uses	The areas of application are similar to those of cinnamon oil. However, cassia oil does not exhibit the same olfactory warmth.

Cascarilla – Croton Species

Cassia – Cinnamomum cassia

Cassie – *Acacia farnesiana*

Name	**Cassie concrète / Cassie absolute**
Botanical Source/ Occurrence	Is obtained from Acacia farnesiana, a small shrub native to the West Indies and belonging to the Leguminosae family. It is cultivated in Southern France and grows wild throughout the whole mediterranean area.
Isolation Procedure	The concrète is obtained via solvent extraction of the flowers. A further extraction of the concrète yields the absolute.
Yield	Approx. 0.5 % as concrète of which some 30 % is obtained as absolute.
Odor	Displays a most unusual combination of both spicy and complex floral notes.
Uses	Is used widely in fine fragrances, because it blends harmoniously with other natural fragrances such as jasmin and rose.

Beaver – *Castor fiber*

Name	**Castoreum**
Source	The beaver, Castor fiber, lives wild in Canada and is raised on farms in U.S.S.R. The secretion is obtained from a gland which is found between the hind legs.
Isolation Procedure	Via extraction of the pods.
Odor	Warm, leather-like odor which is pleasantly sweet upon dilution.
Uses	Is used mostly in the tincture form in fougères, chypres and oriental bases. Displays a very headstrong character which can be difficult to accommodate.

Name	**Cedarleaf oil (Thuja oil)**
Botanical Source/ Occurrence	This oil is obtained from Thuja occidentalis, which grows wild and is also cultivated in North America and Canada.
Isolation Procedure	Via steam distillation of the fresh leaves and twigs.
Yield	0.4–0.6%.
Odor	Displays a very powerful fresh, camphoraceous, pungent odor which is very reminiscent of certain kitchen herbs.
Uses	Combines very well with lavender, pine and citrus notes and is used extensively in masculine fragrances.

Name	**Cedarwood oil**
Botanical Source/ Occurrence	This oil is obtained from the wood of the cedar, Juniperus virginiana, which is native to the U.S.A. The cedarwood oils Texas and Chinese are obtained from the wood of Juniperus mexicana and Cupressus funebris respectively. In addition, cedarwood oil Atlas, is obtained from the wood of the pine tree, Cedrus atlantica, which grows abundantly in North Africa, particularly in Morocco.
Isolation Procedure	By steam distillation of waste wood, such as chippings, saw dust, etc. from saw mills.

Life tree – Thuja occidentalis

Cedar – Juniperus Species

Yield	Approx. 3.5 %, depending upon the quality of the raw material.
Odor	Harmonious, soft, woody odor. Florida oils display a sweet note, whereas Atlas oils are somewhat acidic in character.
Uses	This oil is used extensively in those creations, where warmth and body are required.

Name	**Cedrol $C_{15}H_{26}O$**
Source	Sesquiterpene alcohol. Is isolated from cedarwood oil.
Appearance	White crystals.
Odor	Dry, woody, powdery odor, typical of cedarwood.

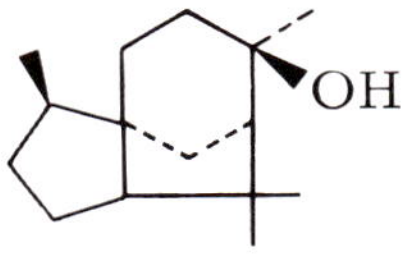

Name	**Cedryl acetate $C_{17}H_{28}O_2$**
Source	Ester. Is synthesized from cedrol and acetic anhydride.
Appearance	Colorless to pale yellow liquid or white crystals.
Odor	Warm, soft, woody odor, typical of cedarwood oil, but also displaying vetiver notes.

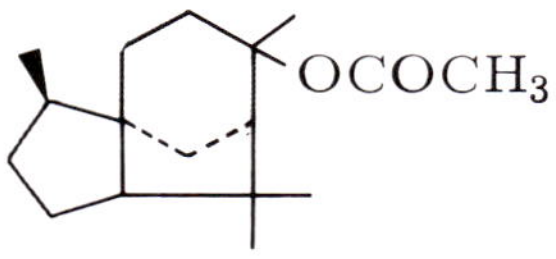

Name	**Celery seed oil**
Botanical Source/ Occurrence	From the celery plant, Apium graveolens. This unbellifer plant is cultivated in Central and Eastern Europe, India and California.
Isolation Procedure	By steam distillation of the crushed seeds.
Yield	Between 2.5 and 3 % depending upon the source of the raw material.
Odor	Warm, spicy, sweet odor, typical of celery.
Uses	Enjoys very wide usage in both, natural type and fantasy fragrances, but due to its diffusiveness and orginality, only at very low levels.

Celery – Apium graveolens

Name	**Chamomile oil blue (German)**
Botanical Source/ Occurrence	This oil is obtained from the common chamomile, Matricaria chamomilla, which is cultivated in Germany, Hungary, Egypt and U.S.S.R.
Isolation Procedure	Via steam distillation of the blossoms.
Yield	0.2 – 0.4 %, depending upon the source of the raw material.
Odor	Typically sweet, herbaceous-like odor with fresh, fruity undertones, somewhat reminiscent of cocoa.
Uses	The oil finds application at low dosages in oriental, floral, chypre and fougère bases.

Name	**Chamomile oil Roman**
Botanical Source/ Occurrence	The oil is obtained from Anthemis nobilis, which is cultivated in England, Bulgaria, Yugoslavia, France and Hungary.
Isolation Procedure	Via steam distillation of the flowers or the whole plant.
Yield	Approx. 1.7 %.
Odor	Fresh, sweet, herbaceous odor with fruity, tea-like undertones.
Uses	Is used very sparingly in perfumes as a toner in floral compositions.

Chamomile German – Matricaria chamomile

Chamomile Roman – Anthemis nobilis

Name	**Champaca concrète / Champaca absolute**
Botanical Source/ Occurrence	This product is obtained from the Champaca tree, Michelia champaca which belongs to the magnolia family. The tree is native to the Philippines and Indonesia.
Isolation Procedure	Extraction of the flowers yields the concrete, from which the absolute is obtained via a second extraction step.
Yield	Very low, usually less than 0.1 %, concrète of which approx. 25 % absolute.
Odor	A heavy, floral, sweet and somewhat herbaceous odor.
Uses	In fine fragrances.

Name	**Cinnamic alcohol $C_9H_{10}O$**
	(structure: benzene ring–CH=CH–CH_2OH)
Source	Alcohol. Can be obtained by the reduction of cinnamic aldehyde. It occurs naturally in hyacinth absolute and styrax.
Appearance	White crystalline mass.
Odor	Warm-balsamic, floral, sweet odor which resembles many types of flowers and to a certain extent, cinnamon oil.

Name	**Cinnamic aldehyde C_9H_8O**
	(structure: benzene ring–CH=CHCHO)
Source	Aldehyde. Can be synthesized from benzaldehyde and acetaldehyde. It is the main component of cinnamon bark and cassia oils.
Appearance	Yellow liquid.
Odor	Powerful odor, very typical of cinnamon.

Name	**Cinnamon bark oil**
Botanical Source/ Occurrence	The oil is obtained from the Ceylonese tree, Cinnamomum ceylanicum Breyne. The cinnamon tree is cultivated, but also grows wild on Sri Lanka, Madagascar, Southern India and the Comoro Islands and the Seychelles.

Cinnamon – Cinnamomum ceylanicum Breyne

Isolation Procedure	The dried bark is subject to steam distillation.
Yield	0.5 – 1 % depending on the source of the raw material.
Odor	Extremely powerful, warm-spicy, sweet odor, typical of cinnamon.
Uses	This oil is a very important component of oriental-type perfumes. When used at very low levels it ellicits a very warm and cosy effect.

Name	**Cinnamon leaf oil**
Botanical Source/ Occurrence	This oil is obtained from the leaves and twigs of Cinnamomum ceylanicum Breyne, the same tree which yields the Ceylon cinnamon bark oil. The tree grows wild and is cultivated in Sri Lanka, South India, Madagascar, the Seychelles and the Comoro Islands.

Cinnamon – Cinnamomum ceylanicum Breyne

Isolation Procedure	By steam distillation of the leaves and twigs.
Yield	1.6 – 1.8 %.
Odor	Warm, spicy odor, typical of cinnamon and cloves.
Uses	Finds application in modern oriental-type perfumes and in fougère and chypre notes.

Name	**Cinnamyl acetate $C_{11}H_{12}O_2$**
	C_6H_5–$CH{=}CHCH_2OCOCH_3$
Source	Ester. Is synthesized from cinnamic alcohol and acetic anhydride. Occurs naturally in cassia and cinnamon oils.
Appearance	Colorless liquid.
Odor	Mild-balsamic, floral-fruity odor with spicy, cinnamon-like undertones.

Name	**Cistus oil s.-c. (Labdanum oil) Labdanum resinoid**
Botanical Source/ Occurrence	From the gum which exudes from the leaves and twigs of Cistus ladaniferus. This shrub grows wild throughout the whole of the mediterranean area, but especially in Spain, Portugal and Southern France.
Isolation Procedure	The resinoid is obtained by extraction and the oil by means of steam distillation.
Yield	Varies considerably depending upon the raw material and the isolation procedure.

Odor	Warm, spicy, ambra-like, balsamic odor with dry, woody notes and considerable diffusivity.
Uses	As an erogenous component in numerous fantasy-type notes.

Name	**Citral $C_{10}H_{16}O$**
	(Geranial) + (Neral)
Source	Aldehyde. A mixture of cis- and trans-isomers. Can be obtained synthetically from isoprene via methyl heptenone and dihydrolinalool as intermediates. Is also obtained from certain essential oils such as lemongrass and Litsea cubeba. Occurs naturally in many other essential oils including lemon, lime and eucalyptus staigeriana oils.
Appearance	Colorless to pale yellow oily liquid.
Odor	Powerful lemon odor with green and bitter notes.

Name	**Citral diethyl acetal $C_{14}H_{26}O_2$**
	$CH(OC_2H_5)_2$
Source	Acetal. Obtained from citral and the triethylester of ortho formic acid.
Appearance	Colorless liquid.
Odor	Bright, fresh lemon-like odor with a bitter, spicy dry-out.

Name	**Citronella oil Java / Citronella oil Ceylon**
Botanical Source/ Occurrence	From two varieties of citronella-grass, Cymbopogon winterianus Jowitt and Cymbopogon nardus, which are cultivated on large plantations in China, Taiwan, Indonesia, Brazil, Columbia and Guatemala.
Isolation Procedure	The dried grass is steam distilled.
Yield	In the region of 1 %.
Odor	Weak, sweet, floral, rose-like odor (Java) or fresh, grassy, camphoraceous odor (Ceylon).
Uses	These oils are used mostly in fragrances designed for household and cleansing products.

Citronella grass – Cymbopogon winterianus Jowitt

Name **Citronellal $C_{10}H_{18}O$**

CHO

Source Aldehyde. Can either be synthesized from pinene via myrcene and geraniol as intermediates or obtained via distillation of citronella or eucalyptus citriodora oils.

Appearance Colorless liquid.

Odor Powerful, herbaceous odor with clear, green citrus notes.

Name **Citronellol $C_{10}H_{20}O$**

CH_2OH

Source Alcohol. Can be synthesized from pinene or obtained by distillation from geranium or citronella oils. Occurs naturally in rose, geranium and citronella oils.

Appearance Colorless liquid.

Odor Fresh, powerful, waxy, rosy odor with sweet-herbaceous undertones.

Name **Citronellyl acetate $C_{12}H_{22}O_2$**

$OCOCH_3$

Source Ester. Is synthesized from citronellol and acetic anhydride. Occurs naturally in citronella oil.

Appearance Colorless liquid.

Odor Fresh, rosy, fruity odor which is light and ethereal.

Name **Civet absolute**

Source Civet is a glandular, paste-like secretion collected from various species of the civet cat, Viverra civetta, which is raised on farms mostly in Ethiopia.

Isolation Procedure The absolute is obtained by extraction of the secretion.

Odor Very powerful, somewhat fecal, animalic odor.

Uses Is included in all high quality fine fragrances.

Civet Cat – Viverra civetta

Name	**Clary Sage absolute / Clary Sage oil**
Botanical Source/ Occurrence	From the clary sage plant, Salvia sclarea, which is cultivated mainly in France, Spain and U.S.S.R.
Isolation Procedure	The concrète is produced by extraction of the flowering tops. The absolute is obtained by further extraction of the concrète, and the oil by steam distillation.
Yield	Approx. 30 % of the concrète is obtained in the form of the absolute. The oil yield is usually less than 1 %.
Odor	Displays a very natural, light, slightly spicy hay-like odor with Bergamot undertones.
Uses	Finds considerable use in fine fragrances, e.g. for fougère bases, chypres and especially Colognes.

Clary Sage – Salvia sclarea

Name	**Clove bud oil**
Botanical Source/ Occurrence	From the clove tree, Eugenia caryophyllata, which is cultivated in Indonesia, Tanzania, Madagascar and Sri Lanka.
Isolation Procedure	This oil is steam distilled from the dried flower buds.
Yield	Approx. 15 %.
Odor	Very powerful, warm, spicy and sweet clove odor.
Uses	Is an important component of many floral fantasy-type perfumes.

Clove – Eugenia caryophyllata Thunb.

Name	**Clove leaf oil**
Botanical Source/ Occurrence	From the clove tree, Eugenia caryophyllata, an evergreen which grows in Indonesia, Tanzania, Madagascar and Sri Lanka.
Isolation Procedure	The oil is steam distilled from the leaves and twigs.
Yield	2–3%.
Odor	Displays a typical clove odor, but is dry and transparent.
Uses	As a spicy, herbaceous component.

Name	**Copaiba balsam**
Botanical Source/ Occurrence	Is a natural oleoresin which occurs as a physiological product in various Copaifera species, for example Copaifera reticulata and Copaifera guyanensis. These large trees grow wild in the North Eastern and Central parts of South America.
Isolation Procedure	Via filtration of the sap.
Yield	Approx. 15 kgs. per tree.
Odor	Mild, pepper-like odor; however, it is lacking in originality.
Uses	Is mainly used for its fixative properties. Improves woody components when used together with patchouli oil.

Copaiba – Copaifera reticulata Ducke

Name	**Coriander oil**
Botanical Source/ Occurrence	Is obtained from Coriandrum sativum, a small herb native to South Eastern Europe. The coriander herb is cultivated in India, U.S.S.R., France, Tunisia, Morocco, Italy and the U.S.A.
Isolation Procedure	The oil is steam distilled from the dried and crushed fully ripe fruits.
Yield	0.8–1.0%.
Odor	Spicy, aromatic odor, typical of the spice and of considerable originality.
Uses	In spicy masculine notes and Colognes where only very small amounts are necessary.

Coriander – Coriandrum sativum

Name **Costus oil**

Botanical Source/ Occurrence: This oil is obtained from Saussurea lappa, a large plant which grows wild in the Himalaya, China and India.

Isolation Procedure: The oil is steam distilled from the comminuted, dried roots.

Yield: Approx. 1 %.

Odor: It has a slightly animalic, rooty, woody odor with erogenous undertones and is very easy to recognize.

Uses: This product is hardly used nowadays and has been successfully replaced by synthetic qualities.

Name **Coumarin $C_9H_6O_2$**

Source: Lactone. Can be synthesized from salicylic aldehyde, acetic anhydride and sodium acetate. Occurs naturally in many plants and essential oils, including tonka beans (Dipteryx odorata), cassia and lavender oils.

Appearance: White crystals or crystalline powder.

Odor: Sweet, herbaceous-warm, somewhat spicy odor which when diluted is reminiscent of freshly cut hay. Displays the typical odor of woodruff.

Name **Cravo oil**

Botanical Source/ Occurrence: Is obtained from a type of tangerine, known in Brazil as Laranja Cravo a member of the Rutaceae family. A hybrid from Citrus aurantium L. and Citrus reticulata.

Isolation Procedure: Cravo oil is machine-pressed from the peel of the ripe fruit.

Yield: 0.5 – 1.2 %.

Odor: Very bright, fresh, fruity odor, somewhere between orange and mandarin.

Uses: Finds considerable use in Colognes and in perfumes with fruity notes.

Cravo – Citrus aurantium × Citrus reticulata

Name **para-Cresol C_7H_8O**

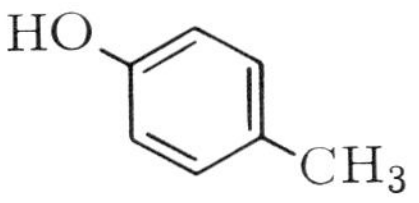

Source Methyl phenol. Can be obtained from para-toluene sulphonic acid by alkali fusion or be isolated from coal tar distillate. Occurs naturally as a trace component in many essential oils.

Appearance White crystals.

Odor Has a smoky, tar-like, medicinal odor which, however, becomes sweet and floral upon dilution.

Uses Useful in very small amounts as a floralizer in artificial ylang-ylang, jasmin and other floral compositions.

Name **para-Cresyl acetate $C_9H_{10}O_2$**

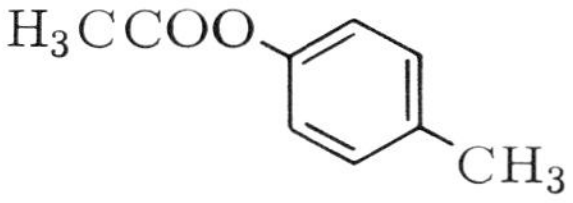

Source Ester. Synthesized via the acetylation of para-cresol.

Appearance Colorless liquid.

Odor Pungent, sweet, urinous odor which upon dilution becomes floral and is reminiscent of narcissus, lily and hyacinth.

Name **para-Cresyl methylether $C_8H_{10}O$**

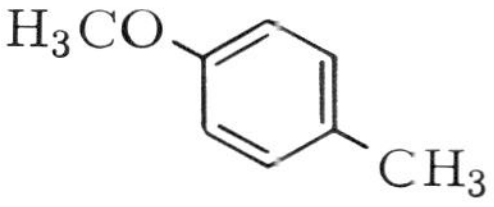

Source Phenol ether. Is synthesized from p-cresol and dimethyl sulphate. Occurs naturally in ylang-ylang and cananga oils.

Appearance Colorless liquid.

Odor Intensely sweet, anisic odor which upon dilution becomes green, fruity and floral, reminiscent of ylang-ylang and narcissus.

Name **Cuminalcohol $C_{10}H_{14}O$**

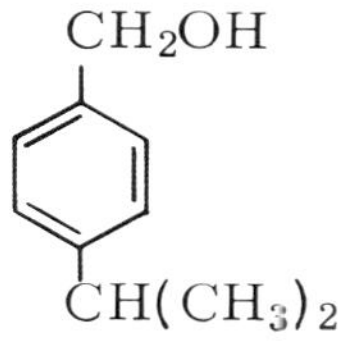

Source Alcohol. Obtained by the reduction of cuminaldehyde.

Appearance Colorless liquid.

Odor Powerful, oily-spicy odor reminiscent of dill seed or caraway.

Name **Cuminaldehyde $C_{10}H_{12}O$**

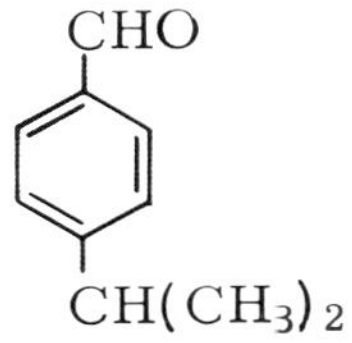

Source Aldehyde. May be synthesized from cumene and carbon monoxide. Is the main component of cumin oil.

Appearance Colorless liquid.

Odor Powerful, pungent, green-herbaceous, spicy-green odor.

Name	**Cumin oil**
Botanical Source/ Occurrence	Is obtained from Cuminum cyminum, a small herb which is cultivated throughout the whole of the mediterranean area and also in India.
Isolation Procedure	Via steam distillation of the dried and crushed seeds.
Yield	2–5 % depending upon production procedure and the source of the raw material.
Odor	Very powerful, soft, green, spicy odor, which is very reminiscent of the spice itself and displays anise-like undertones.
Uses	As a toner with erogenous character. Can only be used in very low dosages.

Name	**Curcuma oil**
Botanical Source/ Occurrence	This product is obtained from the curcuma plant, Curcuma longa, which is related to the ginger plant. It is native to South Asia where it is also cultivated. Main producers are India and Korea.
Isolation Procedure	Via steam distillation of the dried comminuted rhizomes.
Yield	1–5 % depending upon the source of the plant material and the production technique.
Odor	Spicy, fresh, fruity odor.
Uses	Is hardly used any more.

Cumin – Cuminum cyminum

Curcuma – Curcuma longa

Name **Cyclamen aldehyde $C_{13}H_{18}O$**

$(H_3C)_2CH-C_6H_4-CH_2CH(CH_3)CHO$

Source Aldehyde. May be synthesized from cuminaldehyde and propionaldehyde with subsequent hydrogenation.

Appearance Colorless liquid.

Odor Very pleasant, floral-green odor with a watermelon-like note.

Name **Cycloheptadecenone $C_{17}H_{30}O$**

$CH=CH-(CH_2)_7-CO-(CH_2)_7$ (ring)

Source Ketone. Can be synthesized from aleuritic acid. Occurs naturally in the glandular excretion of the civet cat.

Appearance White or colorless crystalline mass.

Odor Mild-sweet, animalic, rather musky odor.

Special Comments Shows excellent fixative properties.

Name **Cyclohexadecenolide $C_{16}H_{28}O_2$**

$CH=CH-(CH_2)_5-CO-O-(CH_2)_8$ (ring)

Source Lactone. Can be synthesized from dihydroxypalmitic acid or from aleuritic acid. It occurs naturally in ambrette seed oil.

Appearance Colorless viscous liquid.

Odor Very pleasant floral, musky odor with a distinctive sandalwood note.

Name **Cyclohexyl oxyacetic acid allylester $C_{11}H_{18}O_3$**

$C_6H_{11}-OCH_2COOCH_2CH=CH_2$

Source Ester. Made from methy phenoxy acetate ester by hydrogenation and subsequent transesterification with allyl alcohol.

Appearance Colorless liquid.

Odor Very strong, somewhat green odor; also fruity, earthy and has a lively galbanum note.

Name **Cyclopentadecanolide $C_{15}H_{28}O_2$**

$(CH_2)_{14}-CO-O$ (ring)

Source Lactone. Belongs to the family of macrocyclic musks. Can be synthesized from ω-hydroxy pentadecanoic acid. It occurs naturally in angelica root oil.

Appearance White, crystalline mass.

Odor Delicate, sweet, musky odor.

Name **Cypress oil**

Botanical Source/Occurrence: Is obtained from the evergreen, Cupressus sempervirens and possibly from other species of Cypressus. The tree grows wild and is also cultivated throughout the whole of the mediterranean area, especially in Algeria and Southern France.

Isolation Procedure: The oil is steam distilled from the leaves and twigs.

Yield: 1.3–1.5 % depending upon the source of the plant material.

Odor: Refreshing, spicy odor, reminiscent of pine-needles and with a limonene note.

Uses: Finds extensive application in modern masculine perfumes.

Cypress – Cupressus sempervirens

D

Name **Davana oil**

Botanical Source/Occurrence: This product is obtained from the flowering herb, Artemisia pallens, which grows wild and is also cultivated in Southern India.

Isolation Procedure: The overground parts of the herb are subject to steam distillation.

Yield: 0.2–0.5 %.

Odor: Sweet, herbaceous odor, somewhat tea-like, reminiscent of dried fruit and displaying a slight whisky note.

Uses: Enjoys only very limited application in fine fragrances.

Name **delta- and gamma-Decalactone**
$C_{10}H_{18}O_2$

Six-membered lactone ring: CH_2, H_2C, CH_2, $H_3C(CH_2)_4$–CH, CO, O

∂-Decalactone

Five-membered lactone ring: H_2C—CH_2, $H_3C(CH_2)_5$–CH, CO, O

γ-Decalactone

Source: Lactones. There are numerous procedures for their synthesis.

Appearance: Colorless, oily liquid.

Odor: Powerful, warm, fruity, nutty odor. gamma-Decalactone exhibits a distinctive peach-note whereas delta-decalactone has a creamy-nutty character.

Name **omega-Decenol $C_{10}H_{20}O$**

$H_2C{=}CH(CH_2)_7CH_2OH$

Source Alcohol. Is synthesized from decamethylene glycol.

Appearance Colorless, oily liquid.

Odor Powerful, fatty-oily, waxy-rosy odor with a fresh and lively character.

Name **n-Decyl acetate $C_{12}H_{24}O_2$**

$H_3CCOO(CH_2)_9CH_3$

Source Ester. Is manufactured from n-decanol and either acetic acid or acetic anhydride.

Appearance Colorless liquid.

Odor Dry, pungent odor of orange-like character with rosy-waxy undertones.

Name **Diacetyl $C_4H_6O_2$**

$H_3CCOCOCH_3$

Source Ketone. Can be obtained by the oxidation of methyl ethyl ketone. Occurs naturally in many essential oils, including angelica root, clove and caraway oils.

Appearance Yellow, oily liquid.

Odor Very powerful butter odor with waxy undertones.

Name **Dihydro citronellol $C_{10}H_{22}O$**

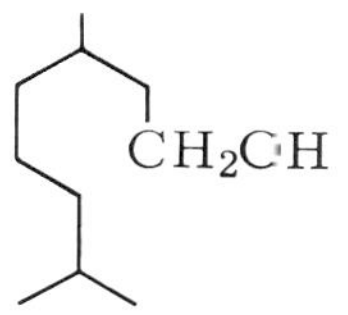

CH_2CH

Source Alcohol. Obtained by the hydrogenation of geraniol, nerol or citronellol.

Appearance Colorless liquid.

Odor Distinctive odor of roses, slightly waxy, somewhat woody and acidic.

Name **Dihydro jasmone $C_{11}H_{18}O$**

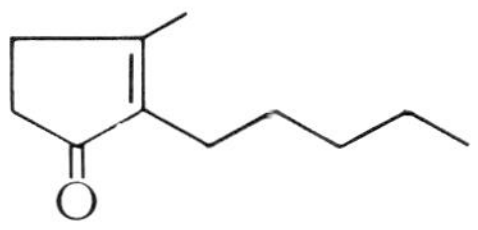

O

Source Ketone. Can be synthesized from heptanal and butenone.

Appearance Colorless liquid.

Odor Intensely floral, jasmin-like odor with fresh and fruity undertones. Is very similar to cis-jasmone in odor, but is not as fine.

Special Comments Should not be mistaken with iso-jasmone or dihydro iso-jasmone, two related isomeric ketones which are used in the manufacture of synthetic jasmin products.

Name **Dihydromyrcenol $C_{10}H_{20}O$**

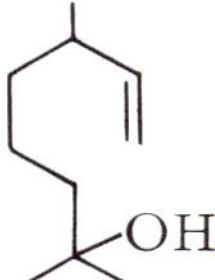

Source Alcohol. Is synthesized either from pinene via dimethyl octadiene as intermediate or by the catalytic hydrogenation of myrcenol.

Appearance Colorless, slightly viscous oil.

Odor Fresh, lively, floral odor with distinct lemon-notes.

Name **Dihydro-nor-dicyclopentadienyl acetate $C_{12}H_{16}O_2$**

Source Ester. Is synthesized through the addition of acetic acid to dicyclopentadiene.

Appearance Colorless or almost colorless liquid.

Odor Powerful herbaceous-green and fresh-woody odor with distinctive fruity-sweet undertones.

Name **Dill weed oil**

Botanical Source/ Occurrence This oil is obtained from the dill plant, Anethum graveolens, which belongs to the umbellifer family. The plant is cultivated in the U.S.A. and the Balkan countries.

Isolation Procedure Via steam distillation of the whole, fully grown herb.

Yield 0.3 – 1.5 % depending upon the source of the plant material.

Dill – Anethum graveolens

Odor Sweet, spicy, very typical odor which is reminiscent of mint.

Uses Is very suitable as a top note in spicy, herbaceous creations.

Name **Dimethylbenzyl carbinol $C_{10}H_{14}O$**

$$C_6H_5-CH_2-C(CH_3)_2-OH$$

Source Alcohol. Is synthesized from benzylmagnesium chloride and acetone via the Grignard reaction.

Appearance White crystalline material.

Odor Warm, herbaceous, floral odor with undertones of freshly cut wood.

Name **Dimethylbenzyl carbinyl acetate $C_{12}H_{16}O_2$**

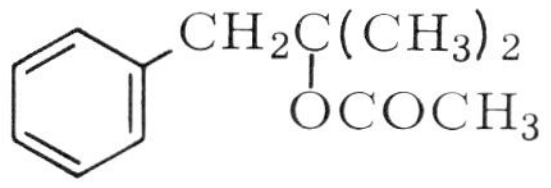

Source Ester. Can be synthesized via the acetylation of dimethylbenzyl carbinol.

Appearance White or colorless crystals.

Odor Fresh, powerful, floral-fruity odor somewhat reminiscent of jasmin, rose and lily.

Name **Dimethylcyclohexene aldehyde $C_9H_{14}O$**

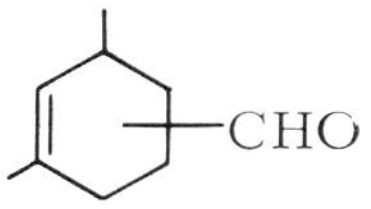

Source Aldehyde. Is obtained from acrolein and dimethyl butadiene via the Diels-Alder reaction. It is composed of two isomers.

Appearance Colorless liquid.

Odor Moderately powerful, sweet, green, grass-like, leafy odor.

Name **Dimethyl heptanol $C_9H_{20}O$**

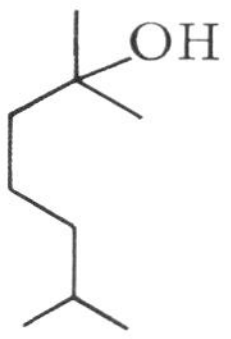

Source Alcohol. Manufactured from methyl heptenone and methyl chloride (Grignard reaction) followed by hydrogenation.

Appearance Colorless liquid.

Odor Very clear, clean and distinctly floral odor, typical of freesia, lily of the valley, etc.

Name **Dipentene $C_{10}H_{16}$**

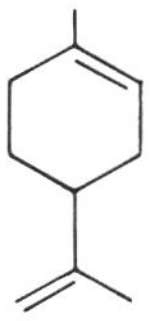

Source Terpene hydrocarbon. Obtained as a by-product in the chemical processing of pinene or turpentine oil.

Appearance Colorless liquid.

Odor A somewhat resinous, slightly sweet odor with overtones reminiscent of pine and fir trees.

Name **Diphenyl methane $C_{13}H_{12}$**

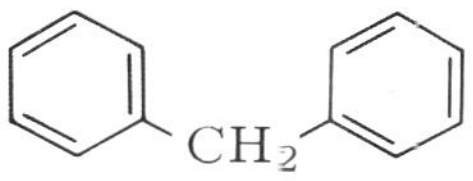

Source Hydrocarbon. Manufactured from benzyl chloride and benzene in the presence of aluminum chloride (Friedel-Crafts reaction).

Appearance Colorless liquid or crystalline mass.

Odor Exhibits a chemical rose odor with a certain sweet, floral note.

Name **Diphenyl oxide $C_{12}H_{10}O$**

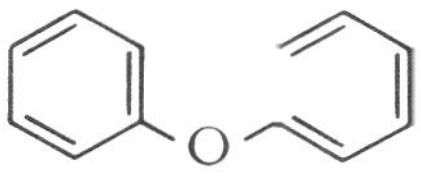

Source Ether. Obtained from chlorobenzene as a by-product in the manufacture of phenol.

Appearance Light yellow liquid or crystals.

Odor Displays a metallic, slightly acidic, rose-like odor.

Name **Dodecahydro-3a,6,6,9a-tetramethyl-(2,1-b)-furan $C_{16}H_{28}O$**

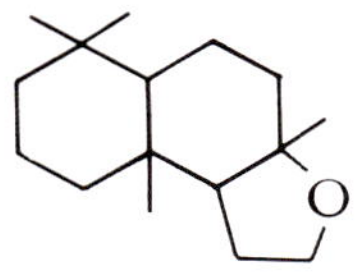

Source Heterocycle. Is synthesized from sclareol which is itself obtained from clary sage.

Appearance Crystalline when pure.

Odor Displays a very intense, warm, ambergris-like odor with woody and orris undertones.

Name **Dwarf pine oil**

Botanical Source/ Occurrence Is obtained from Pinus montana, Pinus mugho and Pinus pumilio which are species of pine-trees native to the European alps.

Isolation Procedure Via steam distillation of the needles and young shoots.

Yield Approx. 1 %.

Odor Typical, fatty-green, pine-needle odor.

Uses In masculine- and conifer-type perfumes.

E

Name **Eau de brouts absolute**

Botanical Source/ Occurrence Is obtained from the bitter orange tree, Citrus bigaradia Risso, which is to be found throughout the whole of the mediterranean area.

Isolation Procedure Via extraction of the distillation waters from numerous essential oils, the most important being neroli oil and petitgrain oil.

Yield Approx. 0.1 %.

Odor Harsh, herbaceous, somewhat bread-like odor of orange blossoms with a soft onion note.

Uses Is used in small amounts in Colognes and fantasy-type perfumes.

Bitter orange – Citrus bigaradia Risso

Name	**Elemi resinoid / Elemi oil**
Botanical Source/ Occurrence	Elemi gum is an exudate from the tree, Canarium luzonicum, which grows wild, but is also cultivated in the Philippines. Both resinoid and oil are obtained from the resinous sap or gum which is produced when the bark of the tree is damaged.
Isolation Procedure	The resinoid is obtained via solvent extraction and the oil by means of steam distillation.
Yield	The yield of resinoid is 80 – 90 % of the crude gum and some 20 – 30 % is obtained in the form of oil.
Odor	Fresh, spicy, lemon-like odor with peppery, green-woody notes.
Uses	Is very useful as a freshener in various perfume compositions, e.g. fougères and conifer types.

Elemi – Canarium luzonicum

Name	**Estragon oil**
Botanical Source/ Occurrence	From the herb, Artemisia dracunculus, which is a member of the Compositae family. The plant is cultivated for culinary purposes in France, Yugoslavia, Spain and Italy.
Isolation Procedure	The whole overground part of the herb – the flowers, stalks and leaves – is steam distilled.
Yield	0.8 – 1.0 %.
Odor	Very typical anise-like, spicy, herbaceous odor which is also reminiscent of celery.
Uses	Is very important as a fresh, spicy component.

Estragon – Artemisia dracunulus

Name **Ethyl acetate $C_4H_8O_2$**

$CH_3COOC_2H_5$

Source Ester. Synthesized from ethyl alcohol and acetic acid.

Appearance Colorless liquid.

Odor Intensely fruity, refreshing, somewhat acidic odor. Very volatile.

Name **Ethyl cinnamate $C_{11}H_{12}O_2$**

$C_6H_5CH=CHCOOC_2H_5$

Source Ester. Can be synthesized from either benzaldehyde and ethyl acetate via the Claisen reaction or from cinnamic acid and ethyl alcohol. Occurs naturally in styrax and Kaempferia galanga.

Appearance Colorless liquid.

Odor Sweet, balsamic-fruity, honey-like odor.

Name **Ethylene brassylate $C_{15}H_{26}O_4$**

CO–O, $(CH_2)_{11}$, $(CH_2)_2$, CO–O (ring)

Source Ester. Is synthesized from ethylene glycol and brassylic acid.

Appearance Colorless liquid.

Odor Sweet, warm, powdery musk-like odor.

Name **Ethyl heptylate $C_9H_{18}O_2$**

$H_3C(CH_2)_5COOC_2H_5$

Source Ester. Obtained by reacting n-heptanoic acid with ethanol.

Appearance Colorless liquid.

Odor Intensely pungent, fruity, sweet odor with a very strong pineapple note.

Name **Ethyl phenylacetate $C_{10}H_{12}O_2$**

$C_6H_5CH_2COOC_2H_5$

Source Ester. Synthesized from phenylacetic acid and ethanol.

Appearance Colorless liquid.

Odor Very intense honey-like odor.

Name **Ethyl vanillin $C_9H_{10}O_3$**

HO, H_5C_2O, CHO (benzene ring)

Source Aldehyde. Can be synthesized from pyrocatechol via guethol.

Appearance White crystals or crystalline powder.

Odor Intensely sweet, creamy, vanilla-like odor.

Name **Eucalyptol (1,8-Cineol) $C_{10}H_{18}O$**

Source Oxide. Occurs in and can be obtained from various essential oils especially the oil of Eucalyptus globulus.

Appearance Colorless liquid.

Odor Very characteristic, fresh, camphoraceous odor, typical of eucalyptus.

Name **Eucalyptus citriodora oil**

Botanical Source/ Occurrence From the tree, Eucalyptus citriodora, which is native to Australia, but is now cultivated mainly in Brazil, South Africa, China and India. The young trees are cut back to a height of about 1.50 metres and thereby develop into bushes.

Isolation Procedure Via steam distillation of the leaves and twigs which are harvested using machines.

Yield The cultivated trees and bushes yield up to 2 % of oil whereas the wild trees, less than 1 %.

Odor Rose-like, grassy odor which is very similar to citronella oil.

Uses Used mainly in perfume compositions for low cost household products. It is an important raw material from which citronellal is obtained.

Eucalyptus – Eucalyptus citriodora

Name **Eucalyptus dives oil**

Botanical Source/ Occurrence From the tree, Eucalyptus dives, which grows wild but is also cultivated in Australia.

Isolation Procedure Via steam distillation of the leaves and twigs.

Yield 2–4 %.

Odor Due to its high content of piperitone and phellandrene, this oil displays a very distinctive mint odor with fresh, grassy undertones.

Uses Used as a raw material for the isolation of piperitone and as a toner in the flavor industry.

Name	**Eucalyptus globulus oil**
Botanical Source/ Occurrence	From the tree, Eucalyptus globulus, which is cultivated in Spain, Portugal and Australia.
Isolation Procedure	By means of steam distillation of the leaves and twigs.
Yield	Between 1.8 and 2 % depending upon the source of the plant material.
Odor	Very powerful, typical odor with a very distinct camphor note.
Uses	Is used extensively in low-cost perfumes for detergents and soaps. Its main area of application is, however, in oral hygiene products.

Eucalyptus – Eucalyptus globulus

Name	**Eucalyptus staigeriana oil**
Botanical Source/ Occurrence	From Eucalyptus staigeriana, a tree native to Australia which, however, is now mainly cultivated in Brazil.
Isolation Procedure	Via steam distillation of the leaves.
Yield	1.9–2.1 %.
Odor	A very distinct, herbaceous, fresh, lemon-like odor.
Uses	In fresh, citric types of perfumes and as a raw material for the isolation of citral.

Name	**Eugenol $C_{10}H_{12}O_2$**
	HO, H_3CO, $CH_2CH{=}CH_2$
Source	Phenol. It is obtained by distilling or extracting either clove leaf oil or clove stem oil which can contain up to 80 % of this material. It is also the main component of clove bud, cinnamon leaf, pimento and bay leaf oils.
Appearance	Pale yellow liquid.
Odor	Warm-spicy, medicinal, rather dry and almost sharp odor reminiscent of cloves.

Name	**Eugenol acetate $C_{12}H_{14}O_3$**
	H_3CCOO, H_3CO, $CH_2CH{=}CH_2$
Source	Ester. Is synthesized by acetylating eugenol. Occurs naturally in clove bud oil.

Appearance White crystals or pale yellow liquid.

Odor Sweet-spicy, balsamic-fruity odor reminiscent of clove.

Name **Eugenol methylether $C_{11}H_{14}O_2$**

H_3CO, H_3CO, $CH_2CH{=}CH_2$

Source Phenol ether. Is synthesized from eugenol and dimethyl sulphate. Occurs naturally in laurel leaf, citronella and pimento oils.

Appearance Colorless liquid.

Odor Herbaceous-spicy, dry-musty odor with slight carnation undertones.

F

Name **Farnesol $C_{15}H_{26}O$**

CH_2OH

Source Sesquiterpene alcohol. Is obtained via the allylic rearrangement of nerolidol. It occurs naturally in cabreuva oil and in musk grains.

Appearance Colorless, oily liquid.

Odor Very delicate, floral fresh-green odor reminiscent of lily of the valley.

Name **Fennel oil bitter / Fennel oil sweet**

Botanical Source/ Occurrence From the seeds of the umbellifers Foeniculum vulgare and Foeniculum vulgare var. dulce, which are cultivated around the Mediterranean.

Isolation Procedure By means of steam distillation of the crushed seeds.

Yield 2–4 %.

Odor Displays a very attractive and typical fennel odor. Sweet fennel oil has a much higher content of anethole.

Uses In herbaceous-type perfumes.

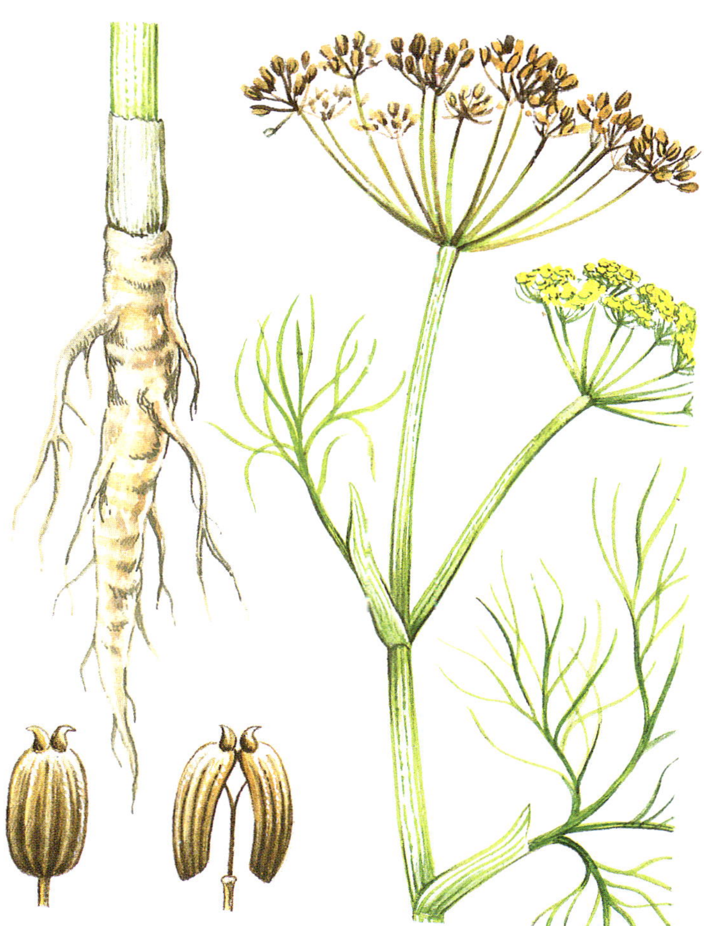

Fennel – Foeniculum vulgare

Name	**Fenugreek resinoid**
Botanical Source/ Occurrence	From Trigonella foenum graecum, a clover-like herb which grows wild and is also cultivated in South Eastern Europe and Asia Minor.
Isolation Procedure	Via solvent extraction of the crushed seeds.
Yield	Varies depending upon the raw material and the isolation procedure.
Odor	Typical walnut odor with woody, bark-like notes.
Uses	In trace quantities as a toner in many different types of perfume compositions.

Fenugreek – Trigonella foenum graecum

Name	**Fir needle oil siberian**
Botanical Source/ Occurrence	From the Siberian fir tree, Abies sibirica Ledeb. This tree however belongs to the pine family.
Isolation Procedure	By means of steam distillation of the young shoots.
Yield	0.8–1.2 %.
Odor	Fresh, powerful pine-forest odor with spicy, fatty undertones.
Uses	Enjoys extensive application in pine-needle-type creations and in masculine notes.

Name	**Flouve oil**
Botanical Source/ Occurrence	From the dried grass, known in France as Flouve odorante (Anthoxanthum odoratum). This grass grows wild in Southern France but is also cultivated there.
Isolation Procedure	Via steam distillation of the grass.
Yield	Varies depending upon the raw material.
Odor	Sweet, herbaceous, hay-like odor which is reminiscent of mimosa.
Uses	At very low levels in many different perfume types including fougères, chypres, oriental bases, etc.

G

Name **Galbanum resinoid / Galbanum oil**

Botanical Source/ Occurrence
From the gum of several different species of Ferula, the main ones being Ferula galbaniflua and Ferula rubricaulis. The gum is obtained from the sap of these large umbellifer plants which dries on contact with air. The plants grow in Iran.

Isolation Procedure
The resinoid is obtained by solvent extraction and the oil via steam distillation.

Yield
The yield of resinoid is 35 – 50 % and some 20 % are obtained as oil.

Odor
Spicy-green, leaf-like odor with woody, pine-needle-like and balsamic undertones. Displays great originality.

Uses
Is used extensively as a green component, especially in floral, herbaceous and conifer-type creations.

Galbanum – Ferula galbaniflua

Name **Geraniol $C_{10}H_{18}O$**

CH_2OH

Source
Alcohol. Can be obtained synthetically from pinene or isolated from natural oils such as palmarosa and citronella oils.

Appearance
Colorless liquid.

Odor
Mild and sweet, floral rose-type odor with dry and bread-like undertones.

Name **Geranium oil**

Botanical Source/ Occurrence
From Pelargonium graveolens and other spices of Pelargonium of which there are approx. 300 different types, hybrids, etc. The plants are cultivated for oil production in Morocco, Egypt, U.S.S.R., France, China, Réunion, Madagascar and Algeria.

Isolation Procedure
By means of steam distillation of the leaves and stalks which are harvested while the plants were in blossom.

Yield
0.1 – 0.2 %, depending upon the plant material.

Odor
The odor varies depending upon the Pelargonium species and where it was grown and produced. For example, geranium oil from Réunion has a full, leaf-like, rose odor with minty-fruity undertones. Geranium oil from North Africa is much lighter and not as minty but displays a stronger rose note.

Uses
This oil is very popular and is used very widely in perfumery.

Geranium – Pelargonium graveolens

Name **Geranyl acetate $C_{12}H_{20}O_2$**

CH_2OCOCH_3

Source Ester. Is synthesized from geraniol and acetic anhydride. Occurs naturally in palmarosa oil and in the oil of Eucalyptus macarthuri.

Appearance Colorless liquid.

Odor Powerful, sweet, fruity-floral and somewhat green odor reminiscent of roses, lavender and pears.

Name **Geranyl formate $C_{11}H_{18}O_2$**

CH_2OCOH

Source Ester. Is synthesized from geraniol and formic acid. It occurs naturally in geranium oil.

Appearance Colorless liquid.

Odor Dry, green-rosy, "leafy" odor with acidic fruity notes.

Name **Geranyl methylether $C_{11}H_{20}O$**

CH_2OCH_3

Source Ester. Is synthesized via the methylation of geraniol.

Appearance Colorless liquid.

Odor Slightly waxy, "leafy" and vegetable-like odor with rose and citrus notes.

Ginger – Zingiber officinale Roscoe

Name	**Ginger oil**
Botanical Source/ Occurrence	From the ginger plant, Zingiber officinale Roscoe, a plant native to India, which is cultivated in Japan, China and India. It also grows wild in Western Africa.
Isolation Procedure	Via steam distillation of the dried, freshly ground rhizomes.
Yield	Approx. 4.4 %, depending upon the isolation procedure and the source of the plant material.
Odor	Warm, spicy-woody odor, reminiscent of citrus and coriander oil with sweet and full undertones. Is very typical of ginger.
Uses	Is often used as a toner in herbaceous notes.

Name	**Gingergrass oil**
Botanical Source/ Occurrence	This oil is obtained from the grass Cymbopogon martini Stapf. var. sofia, which grows wild in India but is also cultivated there. It is related to palmarosa grass, Cymbopogon martini Stapf. var. motia.
Isolation Procedure	Via steam distillation of the fresh grass.
Yield	Approx. 0.2 %.
Odor	Rosy, spicy, herbaceous odor with a slightly sweet, bread-like dry-out.
Uses	This oil is of very little importance in perfumery.

Name	**Grapefruit oil**
Botanical Source/ Occurrence	This oil is obtained from the grapefruit, Citrus decumana, which is cultivated mainly in Israel and the U.S.A.
Isolation Procedure	By the mechanised pressing of the fruit peel.
Yield	Between 0.05 and 0.1 % depending upon the source of the plant material.
Odor	Very bright, fresh, bitter and typical grapefruit odor.
Uses	Is used as a fresh top note in Colognes.

Grapefruit – Citrus decumana

Name	**Guaiacol $C_7H_8O_2$**
	HO, H_3CO (structural formula)
Source	Phenol. Can be synthesized either from ortho-nitrophenol or pyrocatechol. It occurs naturally in neroli oil.

Appearance Colorless crystals.

Odor Powerful, smoke-like, somewhat medicinal, phenolic, bitter odor.

Name **Guaiacwood oil**

Botanical Source/ Occurrence Is obtained from Bulnesia sarmienti, a wild growing tree from the jungles of Paraguay and Argentina.

Isolation Procedure Via steam distillation of the waste wood, branches, saw dust, etc.

Yield Up to 5 % depending upon the isolation procedure.

Odor Balsamic, sweet odor with tea-like and woody, earthy undertones and somewhat fatty.

Uses As a low-cost wood fragrance for household products.

Guaiac – Bulnesia sarmienti

Name **Guaiyl acetate $C_{17}H_{28}O_2$**

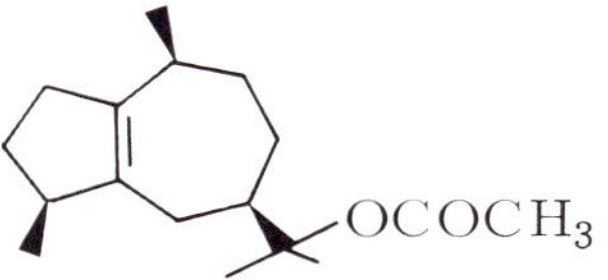

Source Ester. Is obtained by acetylating guaiacwood oil.

Appearance Yellow, slightly viscous liquid.

Odor Sweet-woody, warm, floral odor reminiscent of tea-rose, magnolia and patchouli.

Name **Gurjun balsam / Gurjun balsam oil**

Botanical Source/ Occurrence Gurjun balsam is a natural oleoresin which is exuded from trees of the Dipterocarpus species, mainly Dipterocarpus turbinatus and Dipterocarpus tuberculatus. These grow wild throughout the whole of South Eastern Asia and especially in India.

Isolation Procedure The balsam is obtained by slashing or drilling holes in the trunk of the tree. The oil is produced via steam distillation of the balsam.

Yield One tree can produce about 80 – 100 l. balsam which gives a 60 – 70 % yield of oil.

Odor Displays a relatively weak odor of warm wood, is balsamic, powdery and has a slight patchouli note.

Uses Is used as a fixative.

H

Name — **Helichrysum concrète / Helichrysum absolute**

Botanical Source/ Occurrence — Is obtained from Helichrysum angustifolium, a herb belonging to the composite family which grows wild and is cultivated in the South of France, Hungary, Yugoslavia and several other mediterranean countries.

Isolation Procedure — The concrète is obtained by extraction of the blossoms and further extraction of the concrète gives rise to the absolute.

Yield — Varies depending upon the plant material and isolation procedure.

Odor — Honey-like, sweet, fruity and spicy odor with tobacco and celery-like undertones.

Uses — Is used in herbaceous perfumes.

Name — **Helichrysum oil**

Botanical Source/ Occurrence — From the evergreen herb, Helichrysum angustifolium, which is a composite. The small plant grows wild and is cultivated in France, Italy, Yugoslavia and several other mediterranean countries. It is also known as the Italian everlasting flower.

Isolation Procedure — By means of steam distillation of the herb.

Yield — 0.07 – 0.09 %.

Odor — Powerful, sweet, honey-like, fruity odor with a tea and chamomile note.

Helichrysum – Helichrysum Species

Uses — Is often used in herbaceous perfumes and as a toner in fantasy creations.

Name — **Heliotropin (Piperonal) $C_8H_6O_3$**

Source — Aldehyde. Can be obtained via the oxidation of iso-safrole. It occurs naturally in pepper oil.

Appearance — Colorless or white crystals.

Odor — Sweet, very warm, floral-narcotic odor somewhat reminiscent of almonds.

Name **Heptyl cyclopentanone $C_{12}H_{22}O$**

$(CH_2)_6CH_3$ (structure: 2-substituted cyclopentanone, O)

Source Ketone. Synthesized from cyclopentanone and heptyl aldehyde with subsequent hydrogenation.

Appearance Pale yellow liquid.

Odor Somewhat fruity, green odor reminiscent of jasmin, but also fatty and waxy.

Name **Hexahydro hexamethyl cyclopentabenzopyran $C_{18}H_{26}O$**

Source Is synthesized from pentamethyl indane. Does not occur in nature.

Appearance Colorless liquid.

Odor Sweet, musky odor with fruity undertones.

Name **Hexahydro methyl ionone $C_{14}H_{28}O$**

OH + OH

Source Alcohol. Produced by the hydrogenation of methyl ionone.

Appearance Yellow liquid.

Odor Has a warm, dry, dusty, woody odor with sweet, balsamic undertones, typical of orris.

Name **Hex-2-en-1-al (trans-2-Hexenal) $C_6H_{10}O$**

$H_3CCH_2CH_2CH{=}CHCHO$

Source Aldehyde. Can be synthesized from butylaldehyde and vinyl ethylether.

Appearance Colorless liquid.

Odor Powerful, green-fruity, pungent and fresh odor.

Name **Hex-3-en-1-ol (cis-3-Hexenol) (Leaf alcohol) $C_6H_{12}O$**

$H_3CCH_2CH\overset{c}{=}CHCH_2CH_2OH$

Source Alcohol. Can be synthesized from tetrahydrofuran. Occurs naturally in many essential oils such as geranium, thyme and mullberry leaf oils, but also in violet leaves and tea.

Appearance Colorless liquid.

Odor Very powerful, leafy-green odor reminiscent of freshly cut grass.

Name **Hexenyl acetate $C_8H_{14}O_2$**

$H_3CCOOCH_2CH_2CH\overset{c}{=}CHCH_2CH_3$

$H_3CCOOCH_2CH=CHCH_2CH_2CH_3$

Source Ester. The two isomeric esters, cis-3-hexenyl acetate and trans-3-hexenyl acetate can be synthesized from the corresponding alcohols and acetic acid.

Appearance Colorless liquid.

Odor Powerful, fresh-green, sweet and fruity odor with a distinct note of unripe bananas.

Name **cis-3-Hexenyl salicylate $C_{13}H_{16}O_3$**

$C_6H_4(OH)COOCH_2CH_2CH\overset{c}{=}CHCH_2CH_3$

Source Ester. Synthesized from cis-3-hexenol and salicylic acid.

Appearance Colorless liquid.

Odor Balsamic-sweet, green floral odor.

Name **Hexyl acetate $C_8H_{16}O_2$**

$H_3CCOO(CH_2)_5CH_3$

Source Ester. Synthesized from n-hexanol and acetic acid or acetic anhydride.

Appearance Colorless liquid.

Odor Sweet, fruity-floral odor, somewhat reminiscent of pears with mild, green undertones.

Name **alpha-Hexyl cinnamic aldehyde $C_{15}H_{20}O$**

$C_6H_5CH=C(C_6H_{13})CHO$

Source Can be synthesized from benzaldehyde and octanal.

Appearance Yellow liquid.

Odor Soft, floral odor with distinct jasmin notes.

Name **Hexyl salicylate $C_{13}H_{18}O_3$**

$C_6H_4(OH)COO(CH_2)_5CH_3$

Source Ester. Is synthesized from n-hexanol and salicylic acid.

Appearance Colorless, oily liquid.

Odor Sweet, mild, herbaceous-floral odor with green, bark-like undertones.

Name	**Hop oil**
Botanical Source/ Occurrence	From the hop plant, Humulus lupulus, which is cultivated in Germany and in a number of Eastern European countries.
Isolation Procedure	Via steam distillation of the ripe and fresh flower tops.
Yield	Less than 1 %.
Odor	Harsh, bitter, herbaceous-spicy odor with slight cheese-like undertones.
Uses	Used occasionally in harsh herbal creations.

Name	**Hyacinth absolute**
Botanical Source/ Occurrence	From the hyacinth, Hyacinthus orientalis, which belongs to the lily family and is cultivated on a very large scale in Holland.
Isolation Procedure	Solvent extraction of the blossoms gives the concrète which is subjected to a second extraction step to give the absolute.
Yield	0.01 – 0.02 % as concrète of which some 50 % is obtained as absolute.
Odor	Powerful, sweet, green, leaf-like, floral odor which is reminiscent of styrax and displays a narcotic effect.
Uses	This oil is used very often in green and floral fine fragrances.

Hop – Humulus lupulus

Hyacinth – Hyacinthus orientalis

Name **Hydratropaldehyde dimethyl-acetal $C_{11}H_{16}O_2$**

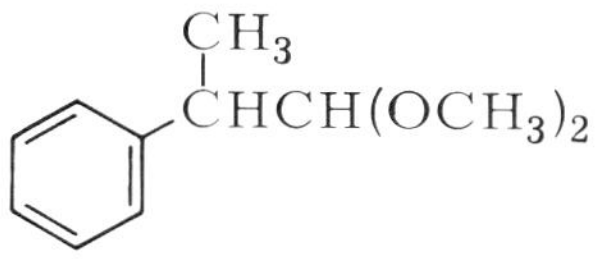

Source Acetal. Synthesized from hydratropaldehyde and methanol.

Appearance Colorless liquid.

Odor Sweet-fruity, spicy-earthy odor reminiscent of mushrooms and walnut.

Name **Hydrocinnamic alcohol (3-Phenyl propan-1-ol) $C_9H_{12}O$**

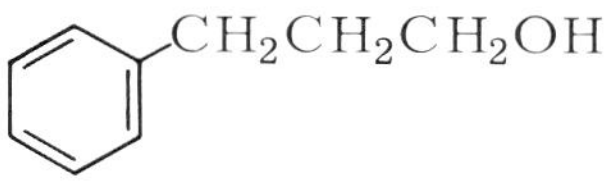

Source Alcohol. Can be synthesized by the catalytic hydrogenation of cinnamic aldehyde. Occurs naturally in styrax resin.

Appearance Colorless liquid.

Odor Warm and mild, balsamic-floral odor with a hyacinth-rose-lilac character.

Name **Hydroquinone dimethylether (Dimethyl hydroquinone) $C_8H_{10}O_2$**

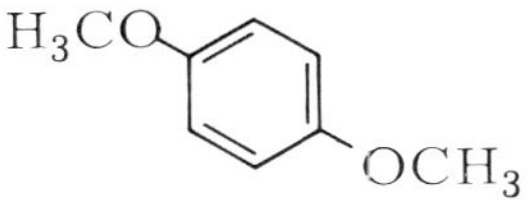

Source Phenolether. Is synthesized from hydroquinone and dimethyl sulphate. Occurs naturally in hyacinth oil.

Appearance White crystals.

Odor Warm-herbaceous, nutty and tobacco-like odor.

Name **para-Hydroxy benzylacetone (Raspberry ketone) $C_{10}H_{12}O_2$**

Source Ketone. Is synthesized from p-hydroxy-benzaldehyde and acetone with subsequent catalytic reduction.

Appearance White crystalline substance.

Odor Very sweet odor, typical of raspberries.

Name **Hydroxycitronellal $C_{10}H_{20}O_2$**

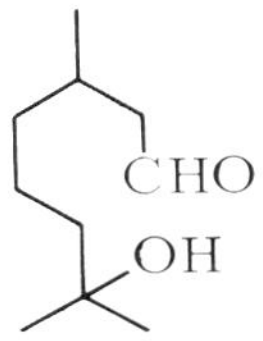

Source Aldehyde. Can be synthesized via the hydration of citronellal.

Appearance Colorless liquid.

Odor Delicate, sweet-floral, elegant odor recalling lily of the valley.

Name **Hydroxycitronellal dimethyl acetal $C_{12}H_{26}O_3$**

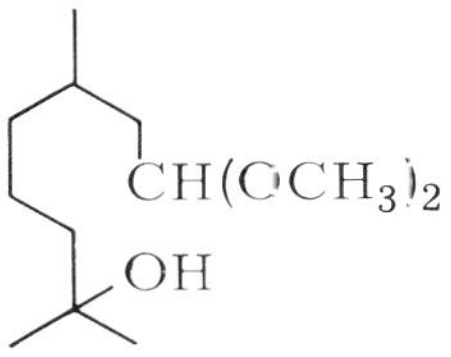

Source Acetal. Synthesized from

hydroxycitronellal and methanol in the presence of hydrogen chloride gas.

Appearance: Colorless, oily liquid.

Odor: Delicate floral odor reminiscent of lily of the valley.

Name: **Hydroxycitronellal methyl-anthranilate $C_{18}H_{27}NO_3$**

CH=N, OH, COOCH₃

Source: Schiff's base. Is synthesized via the condensation of hydroxy-citronellal and methyl anthranilate.

Appearance: Yellow, viscous liquid.

Odor: Very sweet, floral odor reminiscent of orange blossoms and lilies.

Special Comments: It is one of the most widely used Schiff's bases.

Name: **Hydroxy methyl pentyl cyclohexene carboxaldehyde $C_{13}H_{22}O_2$**

OH, CHO

Source: Aldehyde. Can be synthesized via the Diels-Alder condensation of myrcenol and acrolein whereby two isomers are formed.

Appearance: Colorless, viscous liquid.

Odor: Warm, floral odor with distinct lily of the valley notes.

Name: **Hyssop oil**

Botanical Source/ Occurrence: From the culinary herb, Hyssopus officinalis, which grows wild and is also cultivated in many mediterranean countries.

Isolation Procedure: By means of steam distillation of the flowers and stalks of the plant.

Yield: Less than 1 %.

Odor: Sweet, spicy, camphoraceous odor with a warm-woody undertone.

Uses: This oil plays a minor role in perfumery.

Hyssop – Hyssopus officinalis

I

Name **Indole C_8H_7N**

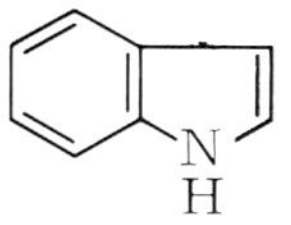

Source Heterocycle. Can be isolated from coal tar. Occurs naturally in neroli oil and jasmin absolute.

Appearance White crystalline substance which turns dark under the influence of light.

Odor Powerful, repulsive, choking odor which, however, when diluted exhibits a floral character highly reminiscent of jasmin and orange blossom.

Name **Ionone $C_{13}H_{20}O$**

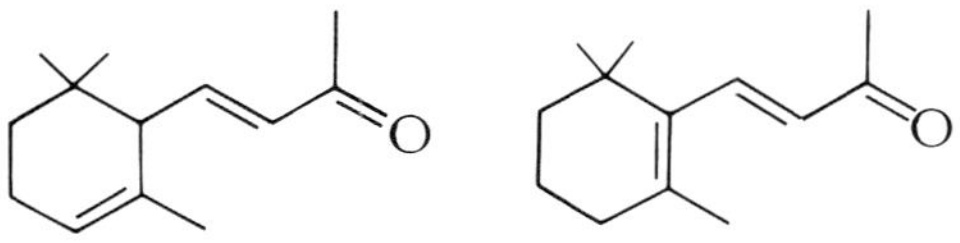

α β

Source Ketone. Consists of the isomers alpha- and beta-ionone. Is synthesized from citral and acetone via pseudo-ionone as intermediate which undergoes cyclization to form ionone. Occurs naturally in boronia absolute.

Appearance Pale yellow liquid.

Odor Is very dependant upon the isomeric ratio. Alpha-ionone displays a classical, sweet, violet-like odor whereas beta-ionone a green-woody, fruity odor which is reminiscent of cedarwood and raspberries. When diluted, the latter becomes very floral with a distinct freesia note.

Special Comments Ionone was synthesized for the first time by Tiemann and Krüger in 1893. It was a by-product in the synthesis of irone, the active odor principle of violets. Nowadays there are many synthetic substances based on the chemical structure of ionone which find use in perfumery. In addition, beta-ionone is one of the starting materials in the synthesis of vitamin A.

Name **alpha-Irone $C_{14}H_{22}O$**

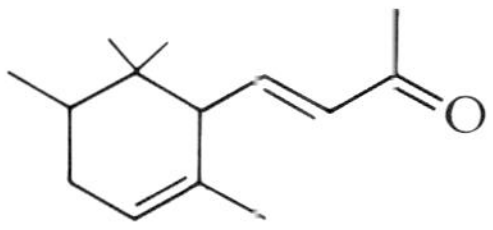

Source Ketone. Can be obtained either by synthesis from dimethyl heptenone via the intermediate pseudo-irone or by isolation from orris root oil.

Appearance Pale yellow to colorless liquid.

Odor Sweet, floral odor reminiscent of violets and orris with warm-woody undertones.

Special Comments Irone is the active odor principle of violets and orris. Synthetic irone is often made up of many isomers. Tiemann and Krüger first isolated irone in 1893 from orris roots. The structural elucidation and synthesis were achieved only many years later by a Swiss research group.

Name **Isoamyl acetate $C_7H_{14}O_2$**

$H_3CCOOCH_2CH_2CH(CH_3)_2$

Source Ester. Is synthesized from isoamyl alcohol and acetic acid. Occurs naturally in Dalmatian sage oil.

Appearance Colorless liquid.

Odor Fresh, pungent, fruity odor, very reminiscent of pears, bananas and raspberries.

Name **Isoamyl cinnamate $C_{14}H_{18}O_2$**

$C_6H_5CH{=}CHCOOCH_2CH_2CH(CH_3)_2$

Source Ester. Is obtained from isoamyl alcohol and cinnamic acid.

Appearance Colorless liquid.

Odor Mild, sweet, balsamic odor somewhat reminiscent of cocoa beans and labdanum.

Name **Isoamyl salicylate $C_{12}H_{16}O_3$**

$C_6H_4(OH)COOCH_2CH_2CH(CH_3)_2$

Source Ester. Is produced from isoamyl alcohol and salicylic acid.

Appearance Colorless liquid.

Odor Sweet, herbaceous, green, mildly floral odor with woody-earthy orchid-like notes.

Name **Isobornyl acetate $C_{12}H_{20}O_2$**

$OCOCH_3$, H

Source Ester. Synthesized from camphene and acetic acid.

Appearance Colorless liquid.

Odor Exhibits the typical character of pine-needle oil, but is a little less natural.

Name **Isobutyl quinoline (sec.-Butyl quinoline) $C_{13}H_{15}N$**

CH_3, HC, H_5C_2, N

Source Heterocyclic compound. Produced from para-sec.-butylaniline.

Appearance Colorless, slightly oily liquid.

Odor Very intense leather odor with slightly acidic, woody and dry undertones. Is a very important component of many chypre creations.

Name **Isobutyl salicylate $C_{11}H_{14}O_3$**

$C_6H_4(OH)COOCH_2CH(CH_3)_2$

Source Ester. Obtained from salicylic acid and isobutanol.

Appearance Colorless liquid.

Odor General floral odor with delicate acidic, metallic and cool notes.

Name **Isocamphyl cyclohexanol (Sandal compound) $C_{16}H_{28}O$**

OH

Source Alcohol. Manufactured via the condensation of camphene and guaiacol with subsequent hydrogenation.

Appearance Colorless, viscous liquid.

Odor Exhibits a very warm, woody, balsamic odor which is very similar to the odor of sandalwood.

Special Comments This product is not a pure compound, but is a mixture of isomers. A multitude of tradenames exist for this synthetic sandal product.

Name **Isoeugenol $C_{10}H_{12}O_2$**

HO, H_3CO, $CH{=}CHCH_3$

Source Phenol. Is obtained via the isomerization of eugenol. Occurs naturally in ylang-ylang and nutmeg oils.

Appearance Pale yellow liquid.

Odor Sweet, very spicy, floral odor, very reminiscent of cloves.

Name **Isoeugenol benzylether $C_{17}H_{18}O_2$**

CH_2O, H_3CO, $CH{=}CHCH_3$

Source Phenol ether. Is isomeric with eugenol benzylether and may be synthesized from benzyl chloride and eugenol with subsequent isomerization.

Appearance White, crystalline material.

Odor Sweet, balsamic, weak clove- and rose-like odor.

Name **Isoeugenol methylether $C_{11}H_{14}O_2$**

H_3CO, H_3CO, $CH{=}CHCH_3$

Source Phenol ether. Is isomeric with eugenol methylether and can be synthesized from isoeugenol and dimethyl sulphate. Occurs naturally in many essential oils, for example citronella oil.

Appearance Colorless liquid.

Odor Mild, sweet, spicy-floral odor recalling cloves and eugenol.

J

Name	**Jasmin absolute**
Botanical Source/ Occurrence	From the jasmin bush, Jasminum grandiflorum, which is native to the East Indies and belongs to the oleander family. The bush is cultivated in Southern France, Spain, Algeria, Morocco, India and Egypt.
Isolation Procedure	Extraction of the flowers gives the concrète which is itself extracted to yield the absolute.
Yield	The yield of concrète is rarely greater than 0.2 %, of which some 50 % is obtained as absolute.
Odor	Powerful, honey-like, sweet, floral odor with fruity-herbaceous undertones.
Uses	Is one of the most important materials used in perfumery and is the most popular flower absolute.

Jasmin – Jasminum grandiflorum

Name	**cis-Jasmone $C_{11}H_{16}O$**
Source	Ketone. May be synthesized from methyl cyclopentenone and pentenyl chloride. Occurs naturally in jasmin absolute.
Appearance	Pale yellow liquid.
Odor	Fruity, celery-like odor which, upon dilution becomes sweet-floral and very reminiscent of jasmin and cherry blossom.

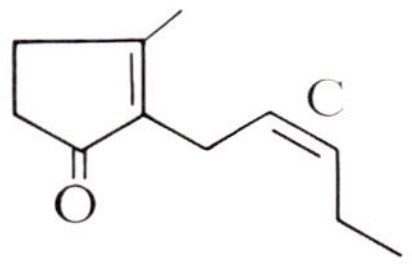

Name	**Juniperberry oil**
Botanical Source/ Occurrence	From the berries of the juniper bush, Juniperus communis, which grows wild all over Europe but mainly in Italy, Yugoslavia, Hungary and Czechoslovakia.
Isolation Procedure	Via steam distillation of the dried, ripe berries.
Yield	0.2 – 2 % depending on the source of the plant material.
Odor	Powerful, green, herbaceous odor with a slight pine-needle undertone. It displays the typical odor of gin.
Uses	Is used as a toner, mostly in masculine notes.

L

Name	**Laurel leaf oil**
Botanical Source/ Occurrence	This oil is obtained from the laurel tree, Laurus nobilis, a small shrub which grows wild but is also cultivated throughout the whole of the mediterranean area.
Isolation Procedure	By means of steam distillation of the fresh, green leaves and twigs.
Yield	1.2 – 1.5 %, depending upon the source of the plant material.
Odor	Fresh, sweet, spicy somewhat camphoraceous odor with a high degree of originality.
Uses	As a spicy note in fantasy creations, especially in masculine perfumes.

Name	**Lavandin oil**
Botanical Source/ Occurrence	Is obtained from the above ground parts of the lavandin plant, which is a hybrid developed by crossing the true lavender plant. Lavandula officinalis, with spike lavender, Lavandula latifolia. The shrub is cultivated in France and Spain.
Isolation Procedure	By means of steam distillation of the dried flowers and stalks.
Yield	1 – 2.5 % depending on the production procedure and the plant material.
Odor	A very typical, refreshing, woody and spicy odor with a distinct camphoraceous note.
Uses	Is used very widely and especially in low-priced products.

Laurel – Laurus nobilis

Juniper – Juniperus communis

Name	**Lavender oil**
Botanical Source/ Occurrence	From Lavandula officinalis or Lavandula vera, a wild growing or cultivated plant, native to the mediterranean countries. It is also cultivated in U.S.S.R., Yugoslavia and Australia.
Isolation Procedure	Via steam distillation of the freshly cut, flowering tops and stalks.
Yield	1.4 – 1.6 %, depending upon the production procedure and plant material.
Odor	Typical, sweet, balsamic, herbaceous odor with floral, woody undertones.
Uses	It is used extensively in Colognes, lavender waters, fougères, chypres, ambres and countless floral perfume types.

Name	**Lemon oil**
Botanical Source/ Occurrence	From the fruit of the lemon tree, Citrus medica, which is cultivated in most mediterranean countries, Brazil, U.S.A., Argentina, Israel and West Africa.
Isolation Procedure	By squeezing the peel of the ripe fruit using special machines.
Yield	0.6 – 0.8 %, depending upon the production procedure.
Odor	A very typical, lively, refreshing odor. It is a classic among the essential oils.
Uses	Universally used. A fresh head note in fantasy perfumes, especially typifies E.d.C. notes.

Lavender – Lavandula officinalis

Lemon – Citrus medica subspec. limonum

Name	**Lemongrass oil**
Botanical Source/ Occurrence	From Cymbopogon flexuosus and Cymbopogon citratus which are cultivated in India, China, Brazil and Guatemala.
Isolation Procedure	Via steam distillation of the fresh or partly dried grass.
Yield	1.8 – 2.2 %, depending upon the source of the plant material.
Odor	Refreshing lemon-like odor, very powerful but somewhat bitter.
Uses	Used in low-cost citrus soap perfumes.

Name	**Lime oil dist. / Lime oil pr.**
Botanical Source/ Occurrence	From the lime, Citrus aurantifolia, which are mainly cultivated for oil production in Brazil, Mexico, Italy and the West Indies.
Isolation Procedure	By steam distillation of the crushed or comminuted fruit or preferably by hand- or machine-pressing the peels of the ripe fruit.
Yield	1.2 – 2.1 % depending upon the production procedure and the source of the raw material.
Odor	Very intense, sweet, lively odor reminiscent of lemon peels.
Uses	Mainly in Colognes as a freshener

Lemongrass – Cymbopogon flexuosus Stapf.

Lime – Citrus aurantifolia Swingle

Name **Limonene $C_{10}H_{16}$**

Source Terpene hydrocarbon. Found in many essential oils from which it can be obtained by distillation.

Appearance Colorless liquid.

Odor Bright, fresh, clean odor typical of citrus fruits such as lemon and orange. Is also somewhat harsh and bitter.

Special Comments Limonene occurs in nature as two optically active forms, d- and l-limonene and also as an optically inactive racemate. d-Limonene is obtained easily since some citrus oils contain up to 90 % of it.

Name **Linalool $C_{10}H_{18}O$**

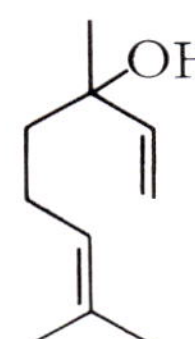

Source Alcohol. Can be obtained from rosewood oil or by synthesis from pinene or methyl heptenone. Occurs naturally in bergamot, lavender and petitgrain oils.

Appearance Colorless liquid.

Odor Displays a very attractive floral odor with slight spicy and lemony undertones.

Name **Litsea cubeba oil**

Botanical Source/ Occurrence From the Litsea cubeba tree which belongs to the laurel family and is native to China and South East Asia.

Isolation Procedure By means of steam distillation of the ripe fruit.

Yield Approx. 2 %.

Odor Very intense, sweet-fruity, lemon-like odor with slightly floral undertones.

Uses Is an important raw material from which citral is obtained and is used widely in citrus perfumes.

Litsea cubeba

Name	**Lovage oil**
Botanical Source/ Occurrence	From the herb Levisticum officinale, an umbellifer which is to be found all over Europe.
Isolation Procedure	By means of steam distillation of both the overground parts of the plant and the fresh, clean roots.
Yield	0.1 – 0.6 % depending upon the plant material.
Odor	Very powerful, sweet-spicy odor reminiscent of celery and angelica root oil but also displaying a weak mossy note.
Uses	Is used occasionally as a toner in spicy, herbal creations.

Lovage – Levisticum officinale Koch

Name	**Linalyl acetate $C_{12}H_{20}O_2$**

OCOCH$_3$

Source	Ester. May be synthesized from dehydrolinalool and acetic anhydride with subsequent catalytic hydrogenation. It occurs naturally in numerous essential oils, especially bergamot, neroli, petitgrain and lavender oils.
Appearance	Colorless liquid.
Odor	Exhibits a radiant, fresh, sweet, floral and fruity odor recalling bergamot and pears.

Name	**Linalyl cinnamate $C_{19}H_{24}O_2$**

OCOCH=CH

Source	Ester. Can be synthesized from cinnamic acid and dehydrolinalool with subsequent catalytic hydrogenation.
Appearance	Colorless to pale yellow liquid.
Odor	Displays a mild, balsamic, herbaceous-floral odor reminiscent of jasmin and lily.

M

Name	**Mace oil / Nutmeg oil**
Botanical Source/ Occurrence	From the nutmeg, Myristica fragrans. The nutmeg tree is cultivated in Indonesia, Sri Lanka and the West Indies.
Isolation Procedure	By steam distillation of the dried, comminuted seeds and peels.
Yield	5 – 15 % depending upon the raw material.
Odor	Very attractive, typical, spicy odor.
Uses	Is used extensively in spicy Colognes and masculine notes.

Name	**Maltol $C_6H_6O_3$**
	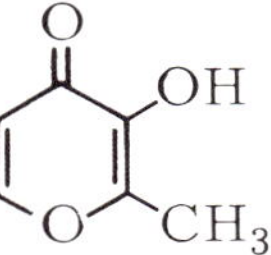
Source	Pyrone and an acid. Obtained mainly from beechwood tar. It occurs naturally in malt and caramel.
Appearance	White crystals.
Odor	Very intense, warm, sweet, fruity and caramel-like odor which smells of strawberries and pineapples when diluted.

Nutmeg – Myristica fragrans

Marjoram – Majorana hortensis Much.

Name	**Marjoram oil**
Botanical Source/ Occurrence	From the plant, Majorana hortensis, which is used as a culinary herb in many countries. It is cultivated for oil production in Germany, France, Spain, Tunisia, Hungary and Egypt.
Isolation Procedure	This oil is steam distilled, both from the fresh and the dried leaves and flowering tops.
Yield	0.5–3 %, depending upon the production procedure and the raw material.
Odor	Displays a typical herbaceous odor.
Uses	Is used as a toner especially in herbal-spicy creations and masculine perfumes.

Name	**Mandarin oil**
Botanical Source/ Occurrence	From Citrus madurensis and Citrus reticulata. The mandarin tree is cultivated in Italy, Brazil, Spain and Argentina.
Isolation Procedure	The peel are cold-pressed by machine.
Yield	0.7–0.8 %.
Odor	Refreshing lively, typical odor with sweet undertones.
Uses	Is a very important component for Colognes and fantasy-type perfumes.

Mandarin – Citrus madurensis Lour.

Mate – Ilex paraguayensis

Name	**Mate absolute**
Botanical Source/ Occurrence	From Ilex paraguayensis, a tree which both grows wild and is also cultivated in most of South America.
Isolation Procedure	Extraction of the leaves gives rise to the concrète which is further extracted to give the absolute.
Yield	Varies depending upon raw material and the isolation procedure.
Odor	Very rich, herbaceous, tea-like odor with hay-like notes.
Uses	In modern fantasy creations.

Name	**Mentha citrata oil**
Botanical Source/ Occurrence	From the mint plant, Mentha citrata, which is a hybrid obtained by crossing Mentha aquatica with Mentha viridis. Is cultivated in the far western United States.
Isolation Procedure	The fully grown plant is steam distilled.
Yield	Approx. 1 %.
Odor	Very powerful, bright, lively, refreshing odor, somewhat resembling bergamot and rosewood oil.
Uses	In fresh notes.

Name	**Menthanyl acetate (Dihydro terpinyl acetate) $C_{12}H_{22}O_2$**
	$OCOCH_3$
Source	Ester. May be synthesized either by the hydrogenation of terpineol and subsequent acetylation or by acetylating first and then hydrogenation.
Appearance	Colorless liquid.
Odor	Fresh, a little acidulous with a weak woody note. Somewhat similar to terpinyl acetate.
Special Comments	This material is not a pure compound, but rather a mixture of isomers.

Name	**l-Menthol $C_{10}H_{20}O$**
	OH
Source	Alcohol. It can be obtained by freezing peppermint oils especially the arvensis types. There are many synthetic procedures starting from, for example thymol, piperitone or citronellal. It occurs naturally in peppermint oils.
Appearance	Colorless crystals.
Odor	It displays a cool, refreshing, sweet and slightly pungent odor which is typically peppermint.
Special Comments	There are 12 isomers of menthol including the racemates.

However, only a few of these are actually used in perfumery. The most commonly used include the naturally occuring l-menthol, its d-isomer and the d,l-racemate. The latter shows a woodier and less sweet odor than l-menthol and is often used as a cheap replacement of the latter.
In addition, l-menthol is a very important flavoring material which finds application in very many foodstuffs.

Name **Menthone $C_{10}H_{18}O$**

Source Ketone. May be obtained by the distillation of peppermint oil or synthetically via the oxidation of l-menthol.

Appearance Colorless liquid.

Odor Exhibits a typical, sweet odor, somewhat reminiscent of menthol.

Special Comments Menthone always refers to l-menthone. However, there are also d-menthone, d- and l-isomenthone and the corresponding racemates.

Name **para-Methyl acetophenone $C_9H_{10}O$**

H_3C – C_6H_4 – $COCH_3$

Source Ketone. Can be synthesized via the Friedel-Crafts reaction from toluene and acetic anhydride in the presence of aluminum chloride. Occurs naturally in rosewood oil.

Appearance Colorless to opaque crystalline substance.

Odor Pungent, almost harsh but warm, floral-sweet odor resembling hawthorn and orange blossom with additional hay-like, bitter almond undertones.

Name **Methyl-n-amylketone $C_7H_{14}O$**

$H_3CCO(CH_2)_4CH_3$

Source Ketone. Can be produced via the hydration of heptyne or from ethyl butyl acetoacetate. Occurs naturally in clove oil and in cinnamon bark oil.

Appearance Colorless liquid.

Odor Penetrating, fruity-spicy "chemical" odor with pear and lavender notes when diluted.

Name **Methyl anthranilate $C_8H_9NO_2$**

NH_2, $COOCH_3$

Source Ester. Produced from anthranilic acid and methanol. Occurs naturally in ylang-ylang and neroli oils as also jasmin absolute.

Appearance Colorless liquid or white to pale yellow crystalline mass.

Odor Medicinal-fruity and somewhat dry-floral acidic odor.

Name **Methyl benzoate (Benzoic acid methylester) $C_8H_8O_2$**

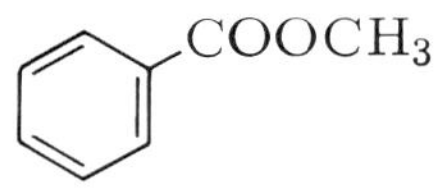

Source Ester. Synthesized from benzoic acid and methanol. It occurs naturally in ylang-ylang oil and tuberose absolute.

Appearance Colorless oil.

Odor Very intense, fruity-floral, black currant-like odor which upon dilution, displays a certain similarity with ylang-ylang and tuberose.

Name **Methyl cinnamate (Cinnamic acid methylester) $C_{10}H_{10}O_2$**

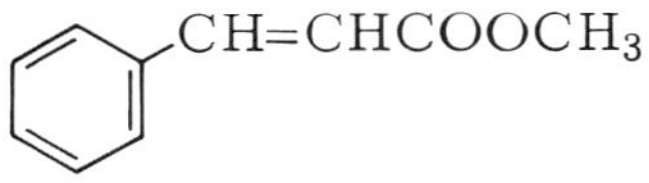

Source Ester. Can be synthesized from cinnamic acid and methanol. Occurs naturally in galanga oil.

Appearance White crystalline material.

Odor Balsamic-fruity odor with slight cinnamon undertones.

Name **Methyl dihydrojasmonate $C_{13}H_{22}O_3$**

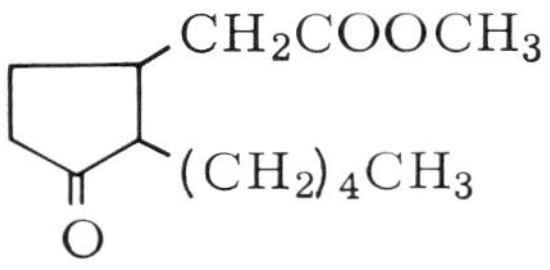

Source Ester. Synthesized from 2-pentyl-cylcopenten-2-one and ethyl malonate. Occurs naturally in tea.

Appearance Pale yellow or colorless liquid.

Odor Very pleasant, sweet-floral jasmin-like odor with fruity undertones.

Name **Methyl heptenone $C_8H_{14}O$**

Source Ketone. Can be synthesized from isoprene and acetone. Methyl heptenone is found in essential oils such as those from lemon-grass and Litsea cubeba.

Appearance Colorless liquid.

Odor Bright, fresh, somewhat acidic odor with sweet, spicy and fruity undertones.

Name **Methyl ionone $C_{14}H_{22}O$**

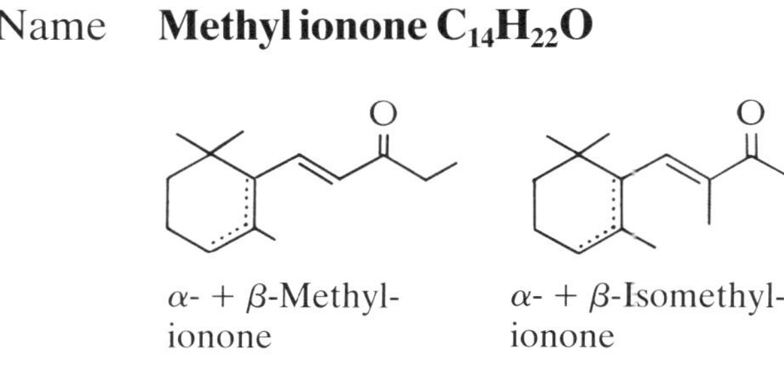

α- + β-Methyl-ionone α- + β-Isomethyl-ionone

Source Ketone. Is synthesized via the condensation of citral with methyl ethyl ketone with subsequent cyclization. The commercial products are often a mixture of isomers.

Appearance Yellow to green liquid.

Odor Generally – displays a sweet, woody, floral odor. alpha-Isomethyl ionone has a creamy, tobacco-like odor whereas beta-isomethyl ionone animalic, ambergris-like notes.

Name **Methyl jasmonate $C_{13}H_{20}O_3$**

CH_2COOCH_3

Source Ester. May be synthesized from muconic acid via the intermediate oxycyclopentenyl acetic acid. It occurs naturally in jasmin absolute.

Appearance Colorless liquid.

Odor Exhibits a soft, floral jasmin odor with weak herbaceous undertones.

Name **Methyl-N-methyl anthranilate (Dimethyl anthranilate) $C_9H_{11}NO_2$**

$NHCH_3$

$COOCH_3$

Source Ester. Is manufactured by the methylation of methyl anthranilate. It is the main component of mandarin-petitgrain oil.

Appearance Colorless to pale yellow liquid.

Odor Displays a musty, floral odor recalling orange blossom and mandarins.

Name **beta-Methyl naphthyl ketone $C_{12}H_{10}O$**

$COCH_3$

Source Ketone. Synthesized from naphthalene and acetylchloride in the presence of aluminum chloride (Friedel-Crafts reaction).

Appearance White crystalline material.

Odor Fruity-floral odor somewhat reminiscent of orange blossom.

Name **Methyl octyl acetaldehyde (MOA) $C_{11}H_{22}O$**

$H_3C(CH_2)_7CH(CH_3)CHO$

Source Aldehyde. Obtained by the reduction of alpha-methylene decanal.

Appearance Colorless liquid.

Odor Exhibits a dry, aldehydic, citrus-like odor which is somewhat reminiscent of orange peels.

Name **Methyl phenylacetate $C_9H_{10}O_2$**

CH_2COOCH_3

Source Ester. Synthesized from phenylacetic acid and methanol.

Appearance Colorless liquid.

Odor Floral, honey-like odor with a slightly pungent, acidic note.

Name: **Methylphenyl glycidic acid ethylester (Strawberry aldehyde / Aldehyde C 16 so-called) $C_{12}H_{14}O_3$**

C_6H_5–$CH(CH_3)$–CH–$COOC_2H_5$

Source: Ester. Synthesized from acetophenone and ethyl monochloroacetate.

Appearance: Colorless liquid.

Odor: Powerful, warm, sweet odor, very reminiscent of strawberries.

Special Comments: This material is an ester and the trivial name "aldehyde C 16" is misleading

---

Name: **Methyl salicylate $C_8H_8O_3$**

$C_6H_4(OH)COOCH_3$

Source: Ester. Is synthesized from salicylic acid and methanol. Occurs naturally in wintergreen oil and in birch bark oil.

Appearance: Colorless liquid.

Odor: Pungent-sweet, rather musty odor with green, medicinal undertones. Reminiscent of wintergreen oil.

Name: **Mimosa concrète / Mimosa absolute**

Botanical Source/ Occurrence: From the mimosa tree, Acacia decurrens, a native of Australia, which is grown in Southern France and Northern Italy (Provence and Piemont).

Isolation Procedure: Mimosa concrète is produced from the flowers and twig ends by extraction. The absolute is prepared from the concrète by a second extraction step.

Yield: Approx. 1 % as concrète, of which some 25 % are obtained as absolute.

Odor: A very powerful, penetrating, green-floral odor with straw-like and bitter undertones.

Uses: In floral fine fragrances.

Mimosa – Acacia decurrens var. dealbata

Name **Muscone (Methyl cyclopentadecanone) $C_{16}H_{30}O$**

CH_3
CH CH_2
$(CH_2)_{12}$
CO

Source Macrocyclic ketone. It can be obtained, for example, from diacetyldodecane via intramolecular condensation with subsequent catalytic reduction. l-Muscone is a component of natural animal musk.

Appearance White or colorless crystalline mass.

Odor Very soft, sweet, musky, erogenous odor.

Name **Musk ambrette $C_{12}H_{16}N_2O_5$**

CH_3
O_2N NO_2
OCH_3
$C(CH_3)_3$

Source Nitro compound. Can be synthesized by alkylating meta-cresol methylether with subsequent nitration.

Appearance Yellow crystals.

Odor Floral-sweet, somewhat heavy, musk-like odor.

Name **Musk ketone $C_{14}H_{18}N_2O_5$**

$COCH_3$
H_3C CH_3
O_2N NO_2
$C(CH_3)_3$

Source Is both a ketone and a nitro compound. Can be manufactured via the Friedel-Crafts acetylation of tertiary-butylxylene and subsequent nitration.

Appearance Pale yellow crystals.

Odor Warm, sweet, erogenous musky odor.

Musk deer – Moschus moschiferus

Name **Musk tincture**

Source The musk deer, Moschus moschiferus is to be found at about 1,500 metres in the mountains of Nepal, China and Tibet. The musk is a glandular secretion contained in an internal pouch between the hind legs of the male animal.

Isolation Procedure The paste-like material is treated with alcohol which extracts the odiferous matter from the secretion. The tincture is then obtained by filtration.

Odor This product displays a typically erogenous, dry, woody, animalic odor.

Uses Nowadays is used only in very expensive fantasy-type perfumes.

Name **Musk xylol $C_{12}H_{15}N_3O_6$**

(Structure: benzene ring substituted with NO_2, CH_3, NO_2, $C(CH_3)_3$, O_2N, H_3C)

Source Nitro compound. Is manufactured by the nitration of alkylated meta-xylene.

Appearance Pale yellow crystals.

Odor Very similar to musk ketone.

Name **Myrrh resinoid / Myrrh oil**

Botanical Source/ Occurrence From the oleoresin-gum exuded by the small trees of various Commiphora species, including Commiphora abyssinica, Commiphora schimperi and Commiphora myrrha. The trees or shrubs grow wild in Somalia, Ethiopia and other North African countries.

Isolation Procedure The resinoid is obtained by extraction of the crude gum; steam distillation of the latter gives the oil.

Yield Depend upon the starting material.

Odor This product displays a very attractive, natural, warm, slightly spicy and sweet odor with balsamic undertones.

Uses The resinoid is used as a fixative and the oil in small amounts in perfumes of the heavy-floral, heavy-oriental type.

Myrrh – Commiphora Species

Name	**Myrtle oil**
Botanical Source/ Occurrence	From Myrtus communis, a small tree or bush which grows throughout the whole of the mediterranean area.
Isolation Procedure	By means of steam distillation of the blossoms, leaves and twigs.
Yield	0.5–1.0 %, depending upon the plant material.
Odor	A very distinct, spicy odor with a camphoraceous and eucalyptol note.
Uses	As a spicy-herbaceous component in many natural-type creations.

N

Name	**Narcissus concrète / Narcissus absolute**
Botanical Source/ Occurrence	From the white narcissus, Narcissus poeticus, which is cultivated in France, Morocco and Egypt.
Isolation Procedure	Extraction of the flowers gives the concrète which is subjected to further extraction to yield the absolute.
Yield	0.2–0.3 % concrète from which the absolute is obtained in approx. 30 % yield.
Odor	The concentrate displays an earthy, hay-like, spicy odor and has to be diluted before becoming typical of the flower.
Uses	In floral fantasy-type perfumes.

Myrtle – Myrtus communis

Narcissus – Narcissus poeticus

Name	**Nerol $C_{10}H_{18}O$**
Source	Alcohol. Can be manufactured along with its isomer, geraniol, from pinene via the intermediate myrcene. It occurs naturally in numerous essential oils such as rose oil and neroli oil.
Appearance	Colorless liquid.
Odor	Displays a fresh, rose-like odor with distinct citrus undertones.

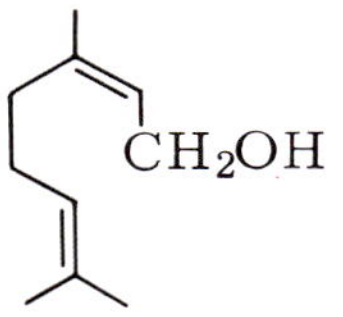

Name	**Neroli oil**
Botanical Source/ Occurrence	From the bitter orange tree, Citrus aurantium subspecies amara and also Citrus bigaradia, which are grown mainly in Morocco, Algeria, Egypt and Southern France.
Isolation Procedure	By means of steam distillation of the flowers which are hand-picked just as they are about to open.
Yield	0.08–0.1 %.
Odor	A very distinct, spicy, bitter odor which is also sweet and displays extreme originality.
Uses	Extensively in Colognes.

Bitter orange – Citrus bigaradia Risso

Name	**Nerolidol $C_{15}H_{26}O$**
Source	Alcohol. Synthesized from linalool via geranyl acetone which is then reacted with acetylene and subsequently partially reduced to yield nerolidol. It is the main component of cabreuva oil from which it is also obtained by distillation.
Appearance	Colorless liquid.
Odor	Very mild, floral, slightly green odor reminiscent of lilies.

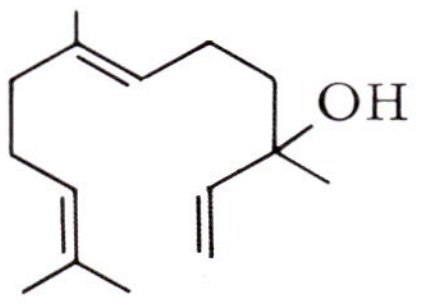

Name **Nerolin Bromelia (beta-Naphthol ethylether) $C_{12}H_{12}O$**

OC_2H_5

Source Ether. Synthesized from beta-naphthol and ethanol in the presence of sulphuric acid.

Appearance White crystals.

Odor Exhibits a heavy, floral, balsamic, sweet odor.

Name **Nerolin Yara Yara (beta-Naphthol methylether) $C_{11}H_{10}O$**

OCH_3

Source Ether. Synthesized from beta-naphthol and dimethyl sulphate.

Appearance White crystals.

Odor Has a warm, sweet, floral odor with narcotic, chemical undertones.

Name **Neryl acetate $C_{12}H_{20}O_2$**

CH_2OCOCH_3

Source Ester. Synthesized from nerol and acetic anhydride. Occurs naturally in helichrysum, neroli and petitgrain oils.

Appearance Colorless liquid.

Odor Floral-fruity, sweet odor somewhat typical of roses, but also displays a slight raspberry note.

Name **2,6-Nonadienal (Violet leaf aldehyde) $C_9H_{14}O$**

$H_3CCH_2CH{=}CH(CH_2)_2CH{=}CHCHO$

Source Aldehyde. Produced by the oxidation of nonadienol. Occurs naturally in violet leaves.

Appearance Colorless to pale yellow liquid.

Odor Very intense, green odor which, when diluted is very reminiscent of violet leaves and cucumbers.

Name **2,6-Nonadienol (Violet leaf alcohol) $C_9H_{16}O$**

$H_3CCH_2CH{=}CH(CH_2)_2CH{=}CHCH_2OH$

Source Alcohol. Synthesized from hexenyl chloride and acrolein via the Grignard reaction with subsequent allyl rearrangement. Occurs naturally in violet leaf oil.

Appearance Colorless liquid.

Odor Extremely intense, sweet, green, herbaceous and vegetable-like odor which displays a very distinct violet and cucumber-like note.

Name **gamma-Nonalactone (Coconut aldehyde / Aldehyde C 18 so-called) $C_9H_{16}O_2$**

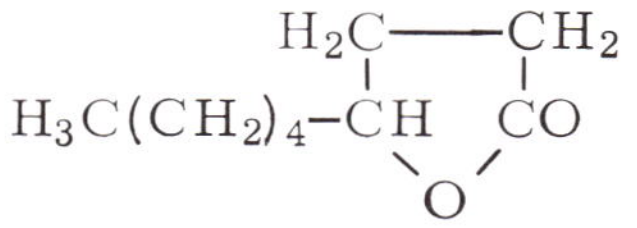

$$\begin{array}{c} H_2C\text{——}CH_2 \\ H_3C(CH_2)_4\text{–}CH \quad CO \\ O \end{array}$$

Source: Lactone. Is synthesized in an analogous manner to undecalactone, for example via the cyclization of nonenoic acid or by reacting n-hexanol with acrylic acid.

Appearance: Colorless liquid.

Odor: Creamy-sweet, soft coconut odor.

Special Comments: This material is a lactone and its historical name "aldehyde C 18" is very misleading.

Name **Nootkatone $C_{15}H_{22}O$**

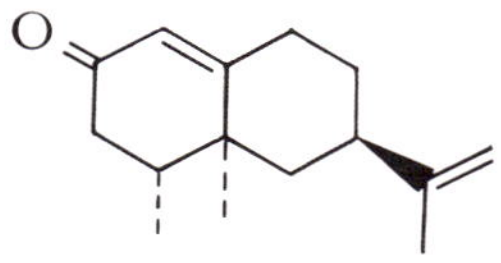

Source: Ketone. Is obtained via the tert.-butyl chromate oxidation of valencene which itself is found in orange oil. Nootkatone is found in grapefruit juice and oil.

Appearance: Pale yellow to colorless liquid.

Odor: Very intense, sweet, fruity, citrus-like odor, typical of grapefruit.

O

Name **Oakmoss resinoid / Oakmoss absolute**

Botanical Source/ Occurrence: From the lichen, Evernia prunastri, which grows primarily on oak trees. It is collected all over Central and Southern Europe, particularly in Yugoslavia and France but also in Morocco.

Isolation Procedure: The resinoid and concrète are obtained by extraction of the moss and the absolute by extracting the resinoid.

Yield: 2–4 % in the form of the resinoid or extract and about half of this as absolute.

Odor: Generally earthy, mossy, spicy, woody odor with slight phenolic and leather-like notes.

Uses: Is used mostly as a fixative.

Oakmoss – Evernia prunastri

Name **Ocimenyl acetate $C_{12}H_{20}O_2$**

OCOCH$_3$

Source Ester. Synthesized via the acetylation of ocimenol.

Appearance Colorless liquid.

Odor Displays a bright, fresh, sweet-herbaceous odor with citrus- and pineapple-like undertones.

Name **Octalactone $C_8H_{14}O_2$**

$$\begin{array}{c} H_2C\text{---}CH_2 \\ H_3C(CH_2)_3\text{-}CH \quad CO \\ O \end{array}$$

Source Lactone. Synthesized from pentanol and acrylic acid.

Appearance Colorless liquid.

Odor Intense coconut odor also reminiscent of coumarin and displays a spicy caraway note.

Name **n-Octyl acetate $C_{10}H_{20}O_2$**

$H_3CCOO(CH_2)_7CH_3$

Source Ester. Synthesized from n-octanol and acetic acid.

Appearance Colorless liquid.

Odor Fruity, floral, waxy odor with aldehydic and slightly acidic undertones.

Name **Olibanum resinoid / Olibanum oil**

Botanical Source/ Occurrence Olibanum is a natural oleoresin which is a physiological product of the tree, Boswellia carterii, which grows wild in Somalia, Ethiopia and in Saudi Arabia. The resinoid and oil are produced from the oleoresin which is formed when the bark of the tree is damaged.

Isolation Procedure The resinoid is obtained via extraction of the oleoresin and the oil by means of steam distillation.

Yield Approx. 50 % resinoid and 10 % as oil.

Odor Balsamic, spicy, slightly lemon-like odor which displays typical incense notes and is somewhat coniferous and resinous.

Uses Is a very important ingredient in many different types of perfumes.

Olibanum – Boswellia Species

Name **Opopanax resinoid / Opopanax oil**

Botanical Source/ Occurrence Opopanax is an oleoresin obtained from the medium-sized tree, Commiphora erythraea var. glabrescens, which grows wild in Somalia and Ethiopia. The resinoid and oil are obtained from the oleoresin which is formed when the bark of the tree is damaged.

Isolation Procedure The resinoid is obtained via extraction of the oleoresin and the oil by means of steam distillation.

Yield Approx. 20 % resinoid and 5–10 % oil.

Odor Very balsamic, warm, sweet odor.

Uses The resinoid is used as a fixative which displays sweet undertones. The oil is used extensively in many different types of perfumes.

Name **Orange flower concrète / Orange flower absolute**

Botanical Source/ Occurrence From the bitter orange tree, Citrus aurantium subspecies amara and also Citrus bigaradia, which are cultivated in most mediterranean countries, especially France, Spain, Algeria and Morocco.

Opopanax – Commiphora erythraea var. glabrescens

Bitter orange – Citrus bigaradia Risso

Isolation Procedure	The concrète is prepared by extraction of the flowers. In order to obtain the absolute, one dissolves the concrète in alcohol and the waxy materials are removed by "freezing out" and filtration. The alcohol is then removed by distillation thereby giving the absolute.
Yield	Approx. 0.3 % concrète from which some 30 % absolute is obtained.
Odor	Very warm, natural, bitter orange odor with spicy undertones and of considerable originality.
Uses	Is used frequently in expensive Colognes and floral fine fragrances.

Bitter orange – Citrus bigaradia Risso

Name	**Orange oil bitter**
Botanical Source/ Occurrence	From the almost ripe fruit of the bitter orange tree, Citrus aurantium subspecies amara and also Citrus bigaradia R., which are cultivated in Morocco, Egypt, Sicily, Spain and France.
Isolation Procedure	The oil is cold-pressed from the peels by hand or by machine.
Yield	0.1 – 0.4 %.
Odor	Displays a typical odor of oranges, is lively, bright, bitter and dry.
Uses	This oil is used extensively to ellicit freshness in top notes.

Name	**Orange oil sweet**
Botanical Source/ Occurrence	From the fruit of the sweet orange tree, Citrus aurantium subspecies dulcis, which is cultivated on large plantations in Brazil, South Africa, the U.S.A. and in most of the mediterranean countries.
Isolation Procedure	By cold-pressing of the fresh peels.
Yield	0.3 – 0.5 % depending upon the raw material.
Odor	Shows a typical odor of the oranges, is lively, fresh, bright and has a fruity sweetness.
Uses	This oil is used extensively, especially in lively citrus-type creations.

Name	**Origanum oil**
Botanical Source/ Occurrence	Origanum oil is produced from the herb, Origanum vulgare, which grows wild in the Middle East and Spain and is also cultivated in France and the Balkans.
Isolation Procedure	The dried, flowering herb is subject to steam distillation.
Yield	1–3 % based on the dry herb. The fresh herb gives a lower yield.
Odor	This oil has an intense phenolic, spicy, bitter odor which is very similar to that of thyme.
Uses	In spicy, natural-type perfumes.

Name	**Orris oil**
Botanical Source/ Occurrence	Iris pallida which is mainly cultivated in and around Florence. Iris germanica and Iris florentina give an oil of poorer quality.
Isolation Procedure	Via the water distillation of the peeled, dried and crushed rhizomes.
Yield	Very low, certainly not more than 0.1 %.
Odor	Oily, woody, violet-like odor with sweet, warm-woody, fruity and floral undertones.
Uses	This product is a very important perfume material which displays a very high degree of originality. It is one of the most expensive materials available to the perfumer.

Origanum – Origanum Species

Orris – Iris pallida

Name	**Osmanthus concrète / Osmanthus absolute**
Botanical Source/ Occurrence	These two products are obtained from the small tree, Osmanthus fragrans, which is native to Japan, China and India.
Isolation Procedure	The concrète is produced by the extraction of the flowers and the absolute by further extraction of the concrète.
Yield	0.1 – 0.2 % concrète of which some 70 % is obtained as absolute.
Odor	These products display a distinct floral odor reminiscent of plums and raisins but with exotic undertones.
Uses	In modern fantasy-type perfumes.

Osmanthus – Osmanthus fragrans

P

Name	**Palmarosa oil**
Botanical Source/ Occurrence	This oil is produced from the grass, Cymbopogon martini Stapf., subspecies motia, which grows wild but is also cultivated in India, Brazil and Central America. The closely related variety, Sofia, is used for producing gingergrass oil which is not so esteemed.
Isolation Procedure	By means of steam distillation of the fresh or dried grass which is harvested just before it flowers.
Yield	Approx. 1.5 %.
Odor	Sweet, floral, rose-like odor with dry grass-like undertones and a bread-like note.
Uses	Used extensively in many perfumes especially in rose notes.

Palmarosagrass – Cymbopogon martini Stapf. var. motia

Name	**Parsley oil seed / Parsley oil herb**
Botanical Source/ Occurrence	From the parsley plant, Petroselinum sativum, an umbellifer plant native to Asia Minor but now found all over the world. It is cultivated in France and the Balkan countries.
Isolation Procedure	By means of steam distillation of the ripe seed or the leaves.
Yield	Approx. 7 % from the seeds but only 0.02 – 3 % from the leaves.
Odor	Warm, herbaceous, spicy odor which displays the typical aroma of the fresh plant.
Uses	As a modifier in herbal-type perfumes.

Parsley – Petroselinum sativum

Name	**Parsnip oil**
Botanical Source/ Occurrence	From the parsnip, Pastinaca sativa, which belongs to the umbellifer family. It grows wild and is also cultivated in Eastern Europe, Central America and South East Asia.
Isolation Procedure	By means of steam distillation of the fruit, flowers and roots.
Yield	Approx. 2 – 3 %, depending on the source of the plant material.
Odor	Aromatic odor, somewhat reminiscent of vetiver and with distinct spicy, herbaceous undertones.
Uses	Is used only very rarely in spicy, herbal-type perfumes.

Parsnip – Pastinaca sativa

Name	**Patchouli oil**
Botanical Source/ Occurrence	From Pogostemon patchouli Pell. and Pogostemon cablin Benth., small shrubs which grow in Indonesia, the Philippines, China and on Madagascar.
Isolation Procedure	The dried and fermented leaves are steam distilled.
Yield	Up to 3 %, depending upon the quality of the leaves.
Odor	Very intense, woody, sweet-balsamic odor with spicy and woody-earthy undertones. Displays extreme originality.
Uses	This oil is one of the most important materials available to a perfumer and is used very widely in many types of perfumes.

Name	**Pennyroyal oil**
Botanical Source/ Occurrence	From the flowering herb, Mentha pulegium, and some other related varieties. It is cultivated in Japan, Spain, Morocco and Tunisia.
Isolation Procedure	The freshly harvested, slightly dried herb is steam distilled.
Yield	1.5–2 %.
Odor	This oil displays a warm, very fresh, minty, somewhat spicy and slightly bitter odor.
Uses	Used occasionally in perfumes, for example in the reproduction of certain essential oils, i.e. geranium.

Patchouli – Pogostemon patchouli Pell.

Pennyroyal – Mentha pulegium

Name	**Pepper oil**
Botanical Source/ Occurrence	From the pepper vine, Piper nigrum, which grows in India, Indonesia, South East Asia and Brazil.
Isolation Procedure	The unripe berries are steam distilled.
Yield	Approx. 2 %.
Odor	Displays the typical, warm, spicy pepper odor with slightly herbaceous undertones.
Uses	Used mainly in masculine perfumes.

Pepper – Piper nigrum

Name	**Peppermint oil Arvensis (Mint oil or Cornmint oil)**
Botanical Source/ Occurrence	From several varieties of the herb, Mentha arvensis, a plant native to Japan and China. It is cultivated mainly in China, Japan, Brazil, Paraguay, Korea and Taiwan.
Isolation Procedure	The essential oil is steam distilled from the dried herb.
Yield	1 – 2 %, depending upon the quality of the herb.
Odor	Powerful, minty, fresh, slightly sweet, herbaceous odor.
Uses	This oil is not used much in perfumery.

Name	**Peppermint oil Piperita**
Botanical Source/ Occurrence	From the herb, Mentha piperita. The English Mitcham plant has now spread throughout the whole world. It is cultivated mainly in Spain, France, Italy, the Balkans and especially in the U.S.A.
Isolation Procedure	Via steam distillation of the partially dried herb which is harvested just as it is about to flower.
Yield	0.1 – 1.0 %, depending upon the source and the production procedure.

Peppermint – Mentha piperita

Odor	Powerful, minty, fresh, grass-like odor with sweet, balsamic undertones.
Uses	The typical peppermint odor is rarely called for in perfumery.

Name	**Peru balsam oil**
Botanical Source/ Occurrence	From Peru balsam, an oleoresin which exudes from the trunk of the large Central American tree, Myroxylon pereirae when injured.
Isolation Procedure	The balsam is distilled to give the oil.
Yield	Approx. 50 %.
Odor	Very full, balsamic, sweet, vanilla-like odor with slight smoky undertones.
Uses	Peru balsam oil is an excellent fixative. All perfume houses have voluntarily decided to use Peru balsam oil instead of the balsam because the latter can sometimes cause irritation of the skin.

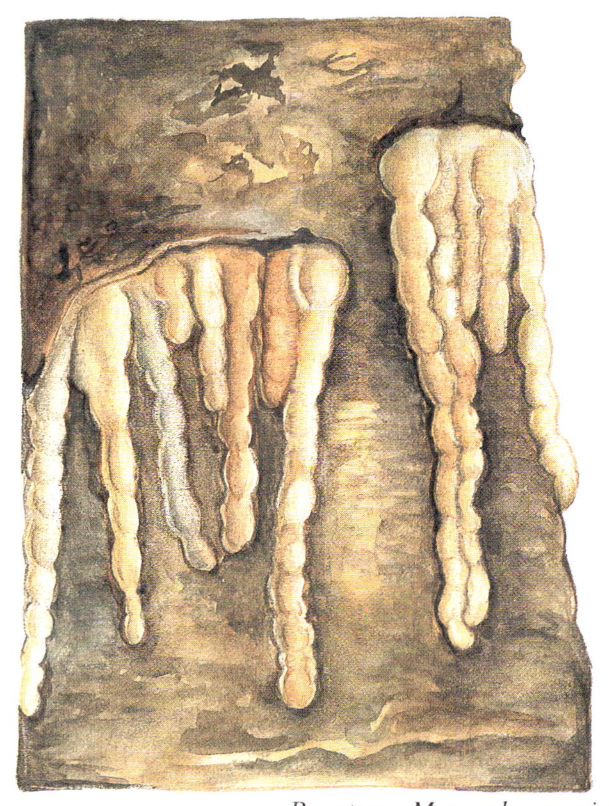

Peru tree – Myroxylon pereirae

Name	**Perilla oil**
Botanical Source/ Occurrence	From the plant, Perilla frutescens, which is cultivated and grows wild in Japan, China and India.
Isolation Procedure	The leaves and flowers are steam distilled.
Yield	0.1 – 0.15 %.
Odor	Very powerful, oily, spicy-green odor with a distinctive cumin note.
Uses	This oil is used only in traces in perfume compositions and has lost some of its original importance.

Perilla – Perilla frutescens

Name	**Petitgrain oil Paraguay**
Botanical Source/ Occurrence	From the bitter orange tree, Citrus aurantium subspecies amara. The trees grow wild but are also cultivated on large plantations in Paraguay, France, Morocco, Algeria, Egypt, Italy and Brazil.
Isolation Procedure	The leaves, twigs and unripe, small fruit are subjected to steam distillation.
Yield	0.5 – 1.0 %.
Odor	Weak, sweet, floral-woody odor reminiscent of neroli oil with bitter, spicy undertones.
Uses	This oil is a very important component of all fresh creations especially Colognes.

Name	**alpha-Phellandrene $C_{10}H_{16}$**
	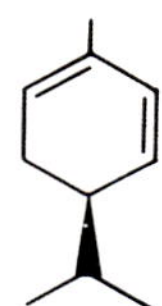
Source	Terpene hydrocarbon. Obtained from the oil of Eucalyptus dives via distillation.
Appearance	Colorless liquid.
Odor	Slightly resinous, fresh odor with earthy undertones reminiscent of pine and spruce forests.

Name **Phenylacetaldehyde (Hyacinth aldehyde) C_8H_8O**

$C_6H_5CH_2CHO$

Source Aldehyde. Synthesized by isomerizing styrene oxide. Occurs naturally in narcissus absolute, neroli and rose oils.

Appearance Colorless liquid.

Odor Very intense, green, floral, hyacinth-like odor with a bitter almond note.

Name **Phenylacetaldehyde dimethyl acetal $C_{10}H_{14}O_2$**

$C_6H_5CH_2CH(OCH_3)_2$

Source Acetal. Produced from phenylacetaldehyde and methanol.

Appearance Colorless liquid.

Odor Intense earthy-green, very distinct leaf-like odor with spicy, floral undertones.

Name **Phenylacetic acid $C_8H_8O_2$**

$C_6H_5CH_2COOH$

Source Acid. Can be obtained from benzyl chloride or benzyl nitrile. Occurs naturally in neroli oil.

Appearance White crystals.

Odor Has a sweet, honey-like odor with a slight animalic note.

Name **Phenylethyl acetate $C_{10}H_{12}O_2$**

$C_6H_5CH_2CH_2OCOCH_3$

Source Ester. Synthesized from phenylethyl alcohol and acetic acid. Occurs naturally in pandanus oil.

Appearance Colorless liquid.

Odor Displays a warm, fruity, rose-like odor with slight sweet and honey-like undertones.

Name **Phenylethyl alcohol $C_8H_{10}O$**

$C_6H_5CH_2CH_2OH$

Source Alcohol. Can be synthesized from benzene and ethylene oxide via the Friedel-Crafts reaction or by the catalytic reduction of styrene oxide. Occurs naturally in rose, geranium and neroli oils.

Appearance Colorless liquid.

Odor Exhibits a mild, warm, rose odor with green, hyacinth undertones.

Name	**Pimento oil**
Botanical Source/ Occurrence	From of the evergreen tree, Pimenta officinalis Lindl., which grows in the West Indies, Central America, India and Réunion.
Isolation Procedure	The oil is steam distilled from the dried, crushed, fully grown but unripe berries or from the leaves.
Yield	3.3–4.5 % from the berries and 0.3–2.9 % from the leaves.
Odor	Balsamic, spicy, pepper-like odor reminiscent of clove oil and with fruity, sweet undertones.
Uses	Used at low dosages in spicy, masculine notes.

Name	**Pine oil**
Botanical Source/ Occurrence	True pine oil is obtained from the heartwood, stumpwood and roots of Pinus palustris, Pinus ponderosa and other Pinus species. The main producers of this oil are the U.S.A., Finland, France, Portugal and U.S.S.R.
Isolation Procedure	The wood chips are steam distilled. Two qualities are available on the market, yellow pine oil and water white oil.
Yield	3–5 %.
Odor	Fresh, bitter, harsh, pine-like odor with resinous and coniferous notes.
Uses	In low-cost perfumes, soap and detergents.

Pimento – Pimenta officinalis Lindl.

Pine – Pinus Species

Name	**Pine needle oil**
Botanical Source/ Occurrence	From the tree, Pinus sylvestris, which grows wild all over Europe and the U.S.S.R.
Isolation Procedure	The needles and twigs are subjected to steam distillation.
Yield	0.1 – 0.5 %.
Odor	Typical fresh, harsh and resinous odor.
Uses	For conifer notes in household products and in masculine perfumes.

Pine – Pinus sylvestris

Name	**Pinene $C_{10}H_{16}$**
Source	Terpene hydrocarbon. Obtained from turpentine oil by distillation. Occurs in very many essential oils.
Appearance	Colorless liquid.
Odor	Very volatile, harsh and spicy odor with fresh undertones, typical of pine and fir trees.
Special Comments	When talking about pinene, one normally means alpha-pinene, but there is also a beta-pinene. Both pinenes are optically active and exist in both the l-, d- and racemic forms.

Name	**Profarnesol $C_{14}H_{26}O$**
	OH
Source	Alcohol. Synthesized from methyl heptenone via a multi-step procedure.
Appearance	Colorless to pale yellow liquid.
Odor	Very delicate, floral odor with balsamic undertones.

R

Name **Resorcinol dimethylether**
$C_8H_{10}O_2$

OCH_3 / OCH_3

Source Phenol ether. Can be synthesized from resorcinol and dimethyl sulphate.

Appearance Colorless liquid.

Odor Powerful, sweet, typical hazelnut-like odor with medicinal, chemical undertones.

Name **Rhodinol**

Source Is a poorly defined mixture of the alcohols, citronellol and geraniol and can be obtained from geranium oil.

Appearance Colorless liquid.

Odor Displays a clear transparent odor of roses and is somewhat sweet, waxy and green.

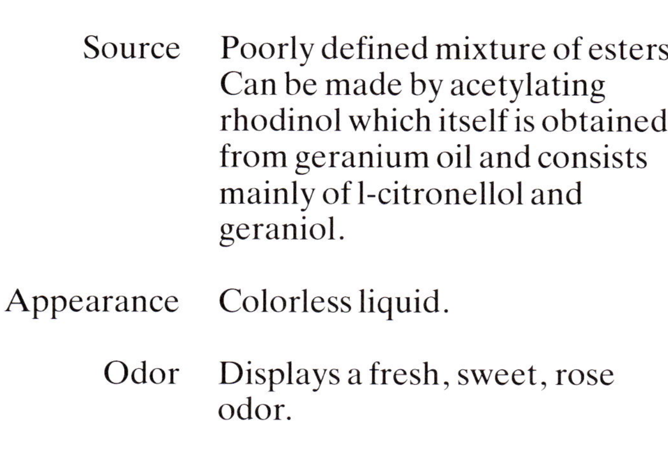

Name **Rhodinyl acetate**

Source Poorly defined mixture of esters. Can be made by acetylating rhodinol which itself is obtained from geranium oil and consists mainly of l-citronellol and geraniol.

Appearance Colorless liquid.

Odor Displays a fresh, sweet, rose odor.

Name **Rose de Mai concrète / Rose de Mai absolute**

Botanical Source/ Occurrence From Rosa centifolia which is to be found all over Europe and even in Asia Minor. The flowers are cultivated in Southern France, Algeria, Morocco and Egypt.

Isolation Procedure The concrète is obtained from the petals by means of enfleurage and further extraction of the concrète gives the absolute.

Yield Approx. 0.25 % as concrète of which some 67 % is obtained as absolute.

Odor Warm, deep, floral rose odor with spicy and honey-like notes.

Uses This oil is a very important component of fine fragrances.

Rose – Rosa centifolia

Name	**Rosemary oil**
Botanical Source/ Occurrence	This oil is obtained from the herb, Rosmarinus officinalis and other related subvarieties which grow mainly in France, Spain, Tunisia and Yugoslavia.
Isolation Procedure	The flowers and leaves are steam distilled.
Yield	1–2 % depending upon the source of the raw material.
Odor	Very powerful, woody, herbaceous odor which is reminiscent of lavender and displays a slight camphoraceous note.
Uses	This product is used extensively, especially in herbaceous types of perfumes.

Name	**Rose oil**
Botanical Source/ Occurrence	From various rose varieties including Rosa centifolia and Rosa damascena. The flowers are cultivated for oil production in France, Italy, Morocco, Bulgaria and Turkey.
Isolation Procedure	The hand-picked petals are steam distilled.
Yield	0.02–0.05 %.
Odor	Very intense and typical rose odor with tea- and honey-like undertones and a soft, green top note.
Uses	Mostly used in fine fragrances.

Rosemary – Rosmarinus officinalis

Rose – Rosa damascena

Name	**Rose oxide $C_{10}H_{18}O$**
Source	Oxide mixture consisting of many isomers. Can be synthesized from citronellol via photo-oxidation with subsequent reduction and cyclization. Occurs naturally in rose and geranium oils.
Appearance	Colorless liquid.
Odor	Displays a very distinctive penetrating, green, grass-like odor and is very reminiscent of rose and geranium.

Name	**Rosewood oil**
Botanical Source/ Occurrence	From the tree, Aniba rosaeodora, which belongs to the laurel family and grows in Central America, Brazil and Mexico.
Isolation Procedure	The chipped wood is steam distilled.
Yield	0.8 – 1.6 %.
Odor	Floral, slightly rose-like odor with spicy, sweet undertones.
Uses	Is used extensively in fantasy-type perfumes and Colognes.

Rose – Aniba rosaeodora

Sage Dalmatian – Salvia officinalis

S

Name	**Sage oil Dalmatian**
Botanical Source/ Occurrence	From the herb, Salvia officinalis, which grows throughout South Eastern Europe, especially Yugoslavia.
Isolation Procedure	The dried leaves are steam distilled.
Yield	1–2 %.
Odor	Powerful, fresh, spicy and herbaceous odor with camphoraceous undertones.
Uses	In herb-type perfumes.

Name	**Sage oil Spanish**
Botanical Source/ Occurrence	From the wild growing herb, Salvia lavandulaefolia. The oil is produced throughout the whole of the Mediterranean and especially in Spain and Southern France where the herb is also cultivated.
Isolation Procedure	The dried herb is steam distilled.
Yield	0.8–1.0 % depending upon the raw material and the production procedure.
Odor	Fresh, herbaceous, camphoraceous, eucalyptus-like odor with slight medicinal undertones.
Uses	In herb-type creations.

Name	**Sandalwood oil East Indian**
Botanical Source/ Occurrence	From the evergreen sandalwood tree, Santalum album, which grows in Indonesia, South East Asia and especially in the Indian province of Mysore.
Isolation Procedure	The coarsely chipped and powdered wood is steam distilled.
Yield	4–6.5 %.
Odor	This oil displays a balsamic, sweet, rich, warm and woody odor with a slight urinous undertone.
Uses	This essential oil is one of the oldest, perfumistically most valuable and most expensive of the raw materials available to the perfumer. It is used to give classic notes in chypres, fougères and oriental bases.

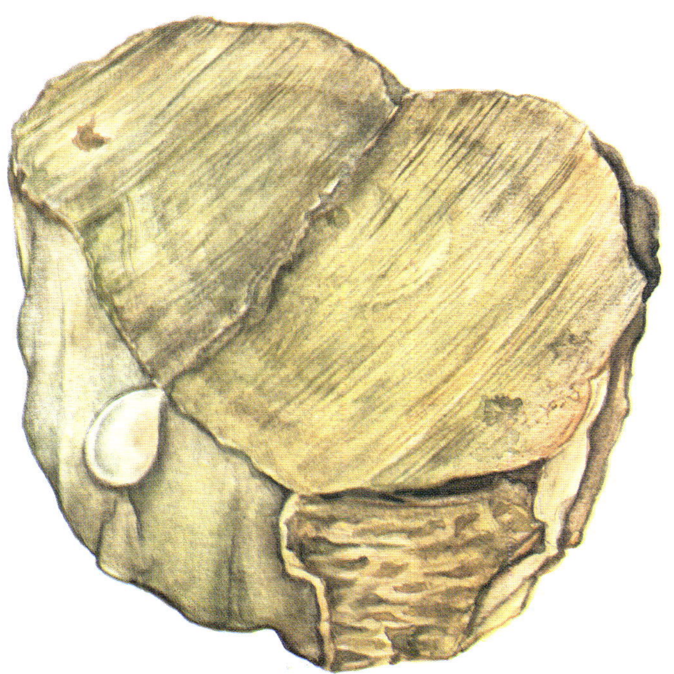

Sandal – Santalum album

Name **Sandalwood oil West Indian so-called (Amyris oil)**

Botanical Source/ Occurrence: This oil is obtained from the tree, Amyris balsamifera, which grows in Indonesia and the West Indies.

Isolation Procedure: The wood chips are steam distilled.

Yield: 1.5 – 3.5 %.

Odor: Mild, woody, sweet, cedarwood-like odor.

Uses: In low-cost, wood-type creations.

Name **Santalol $C_{15}H_{24}O$**

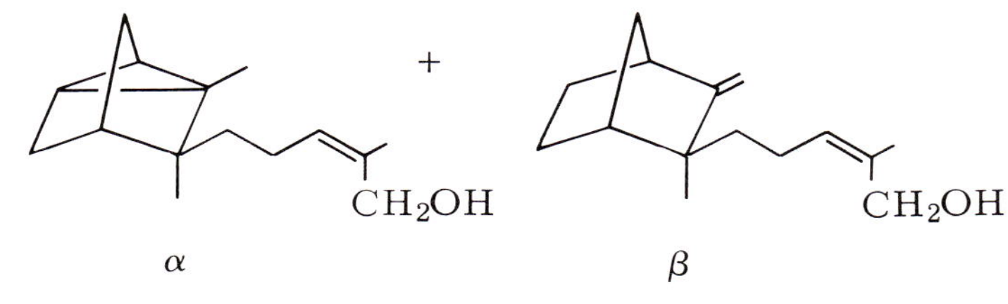

Source: Sesquiterpene alcohol. Is obtained from East Indian sandalwood oil.

Appearance: Colorless liquid.

Odor: Has a very full and typical odor of sandalwood with sweet and woody undertones.

Name **Savory oil**

Botanical Source/ Occurrence: Mainly from the herb, Satureja hortensis, which is cultivated in Yugoslavia, France and the U.S.A.

Isolation Procedure: The twigs and leaves are steam distilled.

Yield: 0.2 – 0.3 %, depending upon the origin of the plant material.

Odor: Fresh, herbaceous, medicinal odor reminiscent of thyme and origanum.

Uses: This product is used at low concentrations in natural herb-type notes.

Savory – Satureja hortensis

Name **Skatole (3-Methylindole)**
C_9H_9N

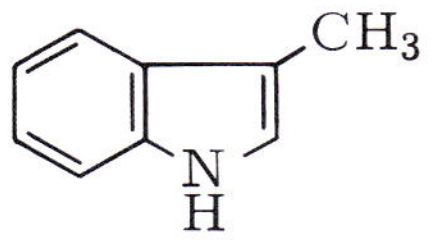

Source Heterocycle. Is obtained from high temperature coal tar. It occurs naturally in the civet absolute.

Appearance White crystals which turn brown when exposed to air and light.

Odor Shows an unpleasant, intensely animalic and medicinal odor. However, on dilution it exhibits a warm, erogenous note which is reminiscent of rotting leaves.

Name **Spearmint oil**

Botanical Source/ Occurrence From the spearmint plant, Mentha spicata and a few related subspecies. The plant is cultivated in China, Brazil, Japan and especially in the U.S.A.

Isolation Procedure The flowering tops of the plant are steam distilled.

Yield Approx. 0.5 %.

Odor Typically green, herbaceous, warm odor.

Uses This product finds only limited use in perfumery. It is mainly used as a flavor in oral hygiene products and chewing gum.

Spearmint – Mentha spicata

Name **Spike Lavender oil**

Botanical Source/ Occurrence From the plant, Lavandula latifolia Vill. (Lavandula spica DC.), which grows wild and is cultivated in the mediterranean countries, especially in Spain, Morocco, Yugoslavia and Southern France.

Isolation Procedure Via steam distillation of the flowering tops of the plant.

Yield 0.5 – 1 % depending upon the source of the plant material.

Odor This oil displays an odor which is very similar to that of lavender but is more herbaceous and camphoraceous.

Uses Used extensively in fougères, masculine perfumes and in lavender-type creations.

Name	**Silver Fir needle oil**
Botanical Source/ Occurrence	From the tree, Abies alba Mill., which belongs to the pine family. It grows wild all over Central Europe and is also known as the silver spruce and white spruce.
Isolation Procedure	The needles and twigs are steam distilled.
Yield	0.2–0.6 %.
Odor	Balsamic-sweet, herbaceous odor somewhat reminiscent of fir needle oil but a little harsher.
Uses	As for fir needle oil.

Name	**Styrax resinoid / Styrax oil**
Botanical Source/ Occurrence	Styrax is a natural balsam formed upon injury of the bark of the Styrax tree, Liquidamber orientalis and Liquidamber styraciflua. The trees grow in Asia Minor, Persia, Honduras and Guatemala.
Isolation Procedure	The resinoid is obtained upon extraction of the balsam followed by neutralization. The oil is produced via steam distillation.
Yield	Approx. 50 % in the form of resinoid and 4–5 % as oil.
Odor	Balsamic-sweet, slightly gas-like odor.
Uses	In many classical fantasy-type perfumes as a fixative.

Silver Fir – Abies alba Mill.

Styrax – Liquidamber Species

Name **Styrolyl acetate (Methyl phenylcarbinyl acetate) $C_{10}H_{12}O_2$**

$C_6H_5-CH(OCOCH_3)CH_3$

Source Ester. Is produced from styrolyl alcohol and acetic acid. Occurs naturally in gardenia absolute.

Appearance Colorless liquid.

Odor Very intense, floral-green, somewhat bitter odor reminiscent of gardenia.

T

Name **Tagetes oil**

Botanical Source/ Occurrence From Tagetes glandulifera and Tagetes patula, the flowers of which have an extremely strong smell. They are cultivated in Italy, Spain and South Africa.

Isolation Procedure The overground parts of the flowering plant are steam distilled.

Yield 0.1–0.5 %.

Odor Very intense, aromatic, herbaceous odor, somewhat reminiscent of fruit and displaying considerable originality.

Uses In extremely floral compositions for fine fragrances and cosmetics.

Tagetes

Name **Thyme linaloe oil**

Botanical Source/ Occurrence From the plant, Thymus quinquecostatus, which grows wild and is also cultivated in China.

Isolation Procedure The whole plant is steam distilled.

Yield Various, dependend on harvest.

Odor Is very similar to rosewood oil but is more spicy and herbaceous.

Uses In fresh citrus-type perfumes.

Name **Thyme oil**

Botanical Source/ Occurrence: Mostly from the herb, Thymus vulgaris, which grows abundantly in France, Spain, Algeria and Yugoslavia.

Isolation Procedure: By means of steam distillation of flowering herb.

Yield: 0.7 – 1 % depending upon production procedure and the raw material.

Odor: Intense, herbaceous, sweet odor of considerable impact and with a medicinal, phenolic note.

Uses: Is an important component in Colognes and herb-type perfumes.

Thyme – Thymus Species

Name **Terpineol $C_{10}H_{18}O$**

OH α

Source: Alcohol. Can either be synthesized from alpha-pinene via terpin hydrate or be obtained by distilling pine oil. Occurs naturally in bergamot, neroli, lime and petitgrain oils.

Appearance: Colorless liquid.

Odor: Floral, sweet odor which is very reminiscent of lilac with slight pine-needle undertones.

Name **Terpinyl acetate $C_{12}H_{20}O_2$**

$OCOCH_3$ α

Source: Ester. Is made from terpineol and acetic anhydride. It occurs naturally in many essential oils, whereby cardamom oil is especially rich in it.

Appearance: Colorless liquid.

Odor: Fresh, sweet, spicy odor with typical citrus undertones.

Name **4,4a,5,9b-Tetrahydroindeno-(1,2:d)-m-dioxin $C_{11}H_{12}O_2$**

Source Cyclic ether. Synthesized from indene and formaldehyde.

Appearance White crystals.

Odor Has an intense animalic, fecal odor with phenolic notes. Exhibits a floral, jasmin-like odor when in dilute solution.

Name **Thymol $C_{10}H_{14}O$**

Source Phenol. Can be obtained via the alkylation of m-cresol with propylene. Thymol is found in many essential oils, the best known of which being thyme oil.

Appearance Colorless crystals.

Odor Intensely spicy odor, typical of thyme.

Name **Tobacco leaf absolute**

Botanical Source/ Occurrence From various species of Nicotiana but mainly from Nicotiana tabacum. The plants are native to America but are now cultivated all over the world.

Isolation Procedure The concrète is obtained by extraction of the leaves and the absolute is produced from the concrète.

Yield Varies depending upon the quality and source of the plant material.

Odor This oil displays a warm, aromatic, sweet, tobacco-like odor of great intensity.

Uses The absolute is essential when genuine tobacco notes are required as for example in certain masculine perfumes.

Tobacco – Nicotiana tabacum

Name	**Tolu balsam resinoid / Tolu balsam oil**
Botanical Source/ Occurrence	From the natural oleoresin produced by the Tolu balsam tree, Myroxylon balsamum. The oleoresin is obtained by slashing the trunks of the trees and hardens upon contract with the air. The trees are to be found in Central and South America.
Isolation Procedure	The resinoid is obtained by extraction and the oil by means of steam distillation.
Yield	Approx. 90 % as resinoid and up to 7 % as oil.
Odor	Balsamic, sweet, hyacinth-like odor of considerable warmth.
Uses	Is used extensively both as a fixative and as a warm, sweet component in floral-type perfumes.

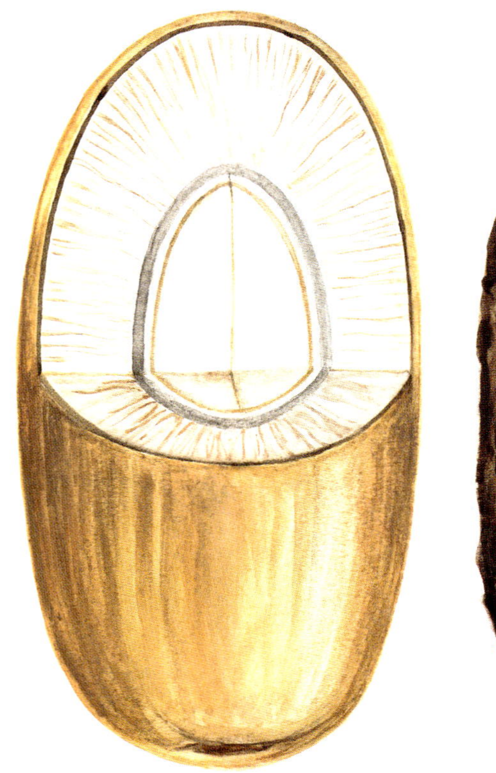

Tonka – Dipteryx odorata

Name	**Tonka bean absolute**
Botanical Source/ Occurrence	From the tonka tree, Dipteryx odorata, which is cultivated in Africa and South America and especially in Brazil.
Isolation Procedure	The dried comminuted beans are extracted to give the concrète from which the absolute is obtained.
Yield	Varies depending upon the production procedure and the plant material.
Odor	Intense, warm, slightly herbaceous, sweet odor which has caramel- and coumarin-like undertones.
Uses	Extensively as a sweet, balsamic fixative in fantasy-type perfumes.

Name	**Treemoss resinoid / Treemoss absolute**
Botanical Source/ Occurrence	From the lichens, Evernia furfuracea and Usnea barbata, which grow on the bark of spruce and fir trees in the humid parts of the forests in Central and Southern Europe. France, Morocco and Yugoslavia are the main producers.
Isolation Procedure	The resinoid is obtained by means of extraction which is then further extracted to yield the absolute.
Yield	2–4 % resinoid depending upon the source of the raw material. Approx. 50 % absolute is obtained based on the resinoid.

Odor: Intense phenolic, tar-like odor which is very powerful and natural.

Uses: Is used as a fixative in many compositions and especially in fougères and chypres.

Treemoss

Name: **Trichloromethyl phenyl carbinyl acetate $C_{10}H_9Cl_3O_2$**

$C_6H_5-CHCCl_3$
$\quad\quad\quad|$
$\quad\quad OCOCH_3$

Source: Ester. Synthesized from benzene and chloral via Friedel-Crafts reaction with subsequent acetylation.

Appearance: White or colorless crystals.

Odor: Mild, floral, balsamic, rose-like odor.

Name: **Trimethyl cyclohexadienyl butenone (Damascenone) $C_{13}H_{18}O$**

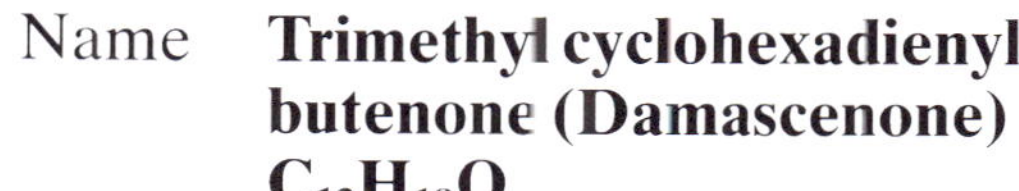

Source: Ketone. May be synthesized via a multi-step procedure, for example from cyclocitral. Damascenone was first found in rose oil.

Appearance: Pale yellow liquid.

Odor: Displays a fruity odor typical of plums and raisins with floral undertones and a bitter background. Is a very important component in modern rose creations

Name **1-(2,6,6-Trimethylcyclohex-2-en-1-yl) but-2-en-1-one (alpha-Damascone) $C_{13}H_{20}O$**

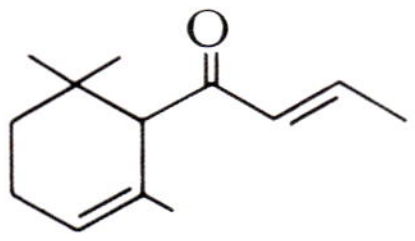

Source Ketone. Can be synthesized from the methyl ester of geranic acid. Occurs naturally in tea aroma.

Appearance Almost colorless liquid.

Odor Exhibits a very intense, floral, fruity odor, very reminiscent of plums, black currants, roses and honey. When diluted it displays a very intense rose odor. By way of comparison, beta-damascone has tobacco and tea-like notes.

Name **Trimethyl undecylenic aldehyde $C_{14}H_{26}O$**

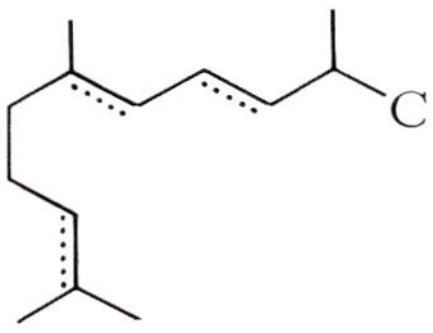

Source Aldehyde. Synthesized from partially reduced pseudo-ionone via the glycidic ester.

Appearance Colorless to pale yellow liquid.

Odor Displays a very intense, sweet, waxy, floral odor which is strongly reminiscent of lily of the valley.

Name **Tuberose concrète / Tuberose absolute**

Botanical Source/ Occurrence From the tuberose plant, Polianthes tuberosa, a native of Central America. Cultivation takes place in France, India, Morocco, Egypt and on the Comoro Islands.

Isolation Procedure The concrète is prepared from the hand-picked flowers by enfleurage and the absolute by means of an additional extraction step.

Yield Approx. 0.1 % concrète from which some 30 % absolute is obtained.

Odor Very heavy, honey-like, sweet, floral odor with narcotic undertones.

Uses Used extensively in modern floral fine fragrances.

Tuberose – Polianthes tuberosa

U

Name	**gamma-Undecalactone (Peach aldehyde/Aldehyde C 14 so-called) $C_{11}H_{20}O_2$**
	$H_3C(CH_2)_6$–CH(–CH_2–CH_2–CO–O–) (ring: H_2C—CH_2, CH, CO, O)
Source	Lactone. Can be synthesized a number of ways, for example via the cyclization of undecenoic acid or through the reaction of n-octanol with acrylic acid.
Appearance	Colorless liquid.
Odor	Displays a very intense, fruity, peach-like odor.

Valerian – Valeriana officinalis

V

Name	**Valerian oil**
Botanical Source/ Occurrence	From the valerian plant, Valeriana officinalis, which grows wild and is also cultivated in Poland, Hungary, France, Belgium, Japan, China and the U.S.S.R.
Isolation Procedure	The dried and comminuted rhizomes are steam distilled.
Yield	Between 0.2 and 2 %, depending upon the quality and age of the plant material.
Odor	Harsh, spicy-balsamic odor which displays considerable originality.
Uses	Can only be used at very low concentrations.

Name	**Vanilla resinoid**
Botanical Source/ Occurrence	From the vanilla plant, Vanilla planifolia, which belongs to the orchid family and is native to Central America. It is cultivated in Indonesia, Madagascar and in the West Indies.
Isolation Procedure	The pods are extracted.
Yield	Varies, depending upon the plant material, the solvent and the extraction.
Odor	This product displays a very typical sweet, warm, balsamic odor of considerable intensity.
Uses	Extensively in fine fragrances in order to produce a sweet-culinary effect.

Vanilla – Vanilla planifolia

Name	**Vanillin $C_8H_8O_3$**

(Structure: HO and H_3CO on benzene ring with CHO)

Source	Phenol aldehyde. Is manufactured from the lignin present in sulphite liquors which are a waste product in the pulp and paper industry. It can also be synthesized from guaiacol and used to be synthesized from eugenol. Occurs naturally in the vanilla pod, in Peru and Tolu balsams and the Siam and Sumatra benzoins.
Appearance	White or cream colored crystals.
Odor	Very intense, sweet, creamy, typical vanilla odor.

Name	**Vassoura oil**
Botanical Source/ Occurrence	From the bush, Baccharis dracunculifolia, which is a native of Brazil.
Isolation Procedure	The fresh, green leaves are steam distilled.
Yield	Approx. 0.3 – 0.5 %.
Odor	Spicy, somewhat grassy odor with woody, erogenous undertones.
Uses	As a harsh, spicy component.

Verbena – Lippia citriodora Kunth

Name **Verbena oil**

Botanical Source/ Occurrence: From the leaves of the approximately 6 feet tall bush Lippia citriodora Kunth, which originates from South America and is now cultivated in Southern France, Algeria and Morocco.

Isolation Procedure: The leaves are subject to steam distillation.

Yield: Less than 1 %.

Odor: Fresh, fruity, citrus-like odor.

Uses: Is not used much nowadays.

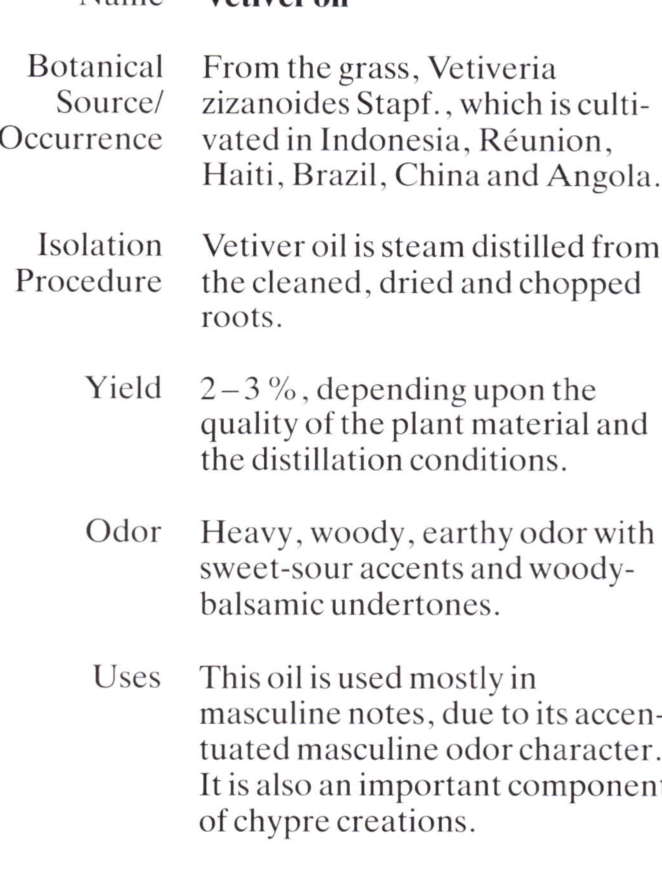

Name **Vetiver oil**

Botanical Source/ Occurrence: From the grass, Vetiveria zizanoides Stapf., which is cultivated in Indonesia, Réunion, Haiti, Brazil, China and Angola.

Isolation Procedure: Vetiver oil is steam distilled from the cleaned, dried and chopped roots.

Yield: 2–3 %, depending upon the quality of the plant material and the distillation conditions.

Odor: Heavy, woody, earthy odor with sweet-sour accents and woody-balsamic undertones.

Uses: This oil is used mostly in masculine notes, due to its accentuated masculine odor character. It is also an important component of chypre creations.

Vetiver – Vetiveria zizanoides Stapf.

Name **Vetiverol**

Source: Mixture of sesquiterpene alcohols which is obtained from various types of vetiver oil.

Appearance: Pale yellow to green liquid.

Odor: Displays a warm, balsamic-earthy odor with woody-dry and sweet undertones.

Name **Vetiveryl acetate**

Source: Ester. Synthesized by the acetylation of vetiver oil with acetic anhydride.

Appearance: Pale to green yellow liquid.

Odor: Displays a fresh, earthy and elegant woody odor.

Name	**Violet leaf concrète / Violet leaf absolute**
Botanical Source/ Occurrence	From the violet, Viola odorata. The plants which are used for the production of perfume material are cultivated mainly in Italy and Southern France.
Isolation Procedure	The concrète is obtained by solvent extraction of the leaves and absolute by further extraction of the concrète.
Yield	Approx. 0.1 % concrète, from which some 30 % absolute is obtained.
Odor	Very intense, green, leafy, herbaceous, peppery odor which displays iris and violet notes.
Uses	Used as a toner and at very low concentrations in special fine fragrances to bring about a green, violet note.

W

Name	**Wormwood oil**
Botanical Source/ Occurrence	From the herb, Artemisia absinthium, a composite plant which grows wild throughout the whole of Europe.
Isolation Procedure	Wormwood oil is steam distilled from the dried leaves and flowering tops of the plant.
Yield	0.5 – 1 %.
Odor	Intense, herbaceous, pungent odor which is reminiscent of cedarleaf oil.
Uses	As a toner. Wormwood oil is a very special product which is used extensively in masculine notes.

Violet – Viola odorata

Wormwood – Artemisia absinthium

Y

Name	**Ylang Ylang oil**
Botanical Source/ Occurrence	From the ylang-ylang tree, Cananga odorata forma genuina, which is native to Indonesia and the Philippines. Cultivation is most extensive on the Comoro Islands. There are four different qualities available, namely extra, first, second and third quality.
Isolation Procedure	The freshly, early-morning picked flowers are steam distilled.
Yield	1.5–2.5%.
Odor	Narcotic, floral, sweet, jasmin-like odor of great diffusion.
Uses	Is a very important oil which generates elegance and warmth in many creations.

Ylang-Ylang – Cananga odorata

Alphabetical Index

A

B

C

Bibliography

Arctander: Perfume and Flavor Materials of natural Origin, Selbstverlag 1960

Arctander: Perfume and Flavor Chemicals, Selbstverlag 1969

Gildemeister/Hoffmann: Die ätherischen Öle, Academy-Publication, Berlin 1966

Guenther: The Essential Oils, Van Nostrand, 1952

Hoppe: Drogenkunde, 8. Edition, de Gruyter, 1975

Müller: Internationaler Kodex der ätherischen Öle, Hüthig Publication 1952/1959

Sukh Dev: Handbook of Terpenoids, CRC Press 1982

Schubert/Wagner: Pflanzennamen und botanische Fachwörter, Neumann Publication 1967

Ullmanns Encyklopädie der technischen Chemie, 3. and 4. Edition, 1963/1981

Uphof: Dictionary of economic plants, J. Cramer Publication 1968

Usher: A Dictionary of plants used by man, Constable London 1974

Zander: Handwörterbuch der Planzennamen, 10. Edition, Eugen Ulmer Publication 1972

Author Index

Nature –
Supplier of Fragrant Perfume Components
Rüdiger Hall

Flowers from the Retort
Dr. Jürgen Nienhaus

Natural and Synthetic Fragrance Components Defined according to: Source, Isolation, Appearance, Yield, Odor, Uses
Rüdiger Hall, Dieter Klemme

We Thank

The Firm Pierre Chauvet S.A., Seillans, France, for editiorial assistance

Allured Publishing, Wheaton, Illinois, USA, for the provision of the graphics from the article “Naturall Essential Oils” by Bernard Meyer-Warnod, Camilli, Albert & Laloue, Grasse, France (Pages 15, 16, 17, 18, 19)

Picture Reference

Photography

Pierre Chauvet S.A., Seillans, France
(Cover)

Peter Gauditz, Hanover, West Germany

Illustrations

Urania-Publication, Leipzig, East Germany

Dorothee Walter, Munich, West Germany

Graphics

Fred Niemüller, Ronnenberg-Benthe, West Germany